Lecture Notes on Multidisciplinary Industrial Engineering

Series Editor

J. Paulo Davim, Department of Mechanical Engineering, University of Aveiro, Aveiro, Portugal

"Lecture Notes on Multidisciplinary Industrial Engineering" publishes special volumes of conferences, workshops and symposia in interdisciplinary topics of interest. Disciplines such as materials science, nanosciences, sustainability science, management sciences, computational sciences, mechanical engineering, industrial engineering, manufacturing, mechatronics, electrical engineering, environmental and civil engineering, chemical engineering, systems engineering and biomedical engineering are covered. Selected and peer-reviewed papers from events in these fields can be considered for publication in this series.

More information about this series at http://www.springer.com/series/15734

Inderdeep Singh · Pramendra Kumar Bajpai ·
Kuldeep Panwar
Editors

Trends in Materials Engineering

Select Proceedings of ICFTMM 2018

Springer

Editors
Inderdeep Singh
Department of Mechanical and Industrial Engineering
Indian Institute of Technology Roorkee
Roorkee, Uttarakhand, India

Pramendra Kumar Bajpai
Netaji Subhas University of Technology
New Delhi, Delhi, India

Kuldeep Panwar
Mechanical and Automation Engineering,
Delhi Technical Campus
Guru Gobind Singh Indraprastha University
Greater Noida, Uttar Pradesh, India

ISSN 2522-5022 ISSN 2522-5030 (electronic)
Lecture Notes on Multidisciplinary Industrial Engineering
ISBN 978-981-13-9015-9 ISBN 978-981-13-9016-6 (eBook)
https://doi.org/10.1007/978-981-13-9016-6

This Springer imprint is published by the registered company Springer Nature Singapore Pte Ltd.
The registered company address is: 152 Beach Road, #21-01/04 Gateway East, Singapore 189721, Singapore

Preface

This book covers a variety of topics from the diverse field of materials. The content of the edited book is based on papers presented at the 1st International Conference on Futuristic Trends in Materials and Manufacturing (ICFTMM-2018) held in Delhi Technical Campus, 28/1, Knowledge Park-III, Greater Noida, Uttar Pradesh 201306, India, on October 26–27, 2018. The topics include research findings on smart materials, advanced metals and alloys, polymeric materials, and composites. These materials are now used in a diverse range of industries, including automotive, marine, aerospace, sporting goods, infrastructure, and others. This book brings together engineers, material scientists, technologists, and researchers working in the field of materials.

This book covers in depth characterization of different advanced materials. It will also offer an insight into the innovation of cost-effective and high-performance materials products and offers a strong introduction to fundamental concepts on the basis of materials science. It conveys the central issue of materials science. It is expected that this book will be suitable for the students, academicians, engineers, and researchers working in this field.

At the end of this preface, I would like to thank Prof. Asanga Ratnaweera, Mechanical Engineering, University of Peradeniya, Sri Lanka, and Prof. Dharam Buddhi, Vice Chancellor, Suresh Gyan Vihar University, India, for their keynote speech in the conference.

Last but not least, support from Management Delhi Technical Campus, Greater Noida, was absolutely essential for the conference.

Roorkee, India	Inderdeep Singh
New Delhi, India	Pramendra Kumar Bajpai
Greater Noida, India	Kuldeep Panwar

Contents

About the Editors

Dr. Inderdeep Singh completed his Bachelor of Technology (Mechanical Engineering) in 1998 from the National Institute of Technology (erstwhile REC) Hamirpur with Honors. He completed his Masters and Ph.D. from Department of Mechanical Engineering, Indian Institute of Technology (IIT) Delhi in 2000 and 2004, respectively. His doctoral research work was awarded by the Foundation for Innovation and Technology Transfer (FITT) as the 'Best Industry Relevant Ph.D. Project' during the session 2004–05. Dr. Singh is currently Associate Professor in the Department of Mechanical and Industrial Engineering, Indian Institute of Technology (IIT) Roorkee. He has published more than 180 articles in international and national journals and conferences. He has authored a book titled *Primary and Secondary Manufacturing of Polymer Matrix Composites* and has also contributed 15 book chapters in the broad area of composite materials. His current research focus is towards developing fully biodegradable green composites.

Dr. Pramendra Kumar Bajpai is an Assistant Professor in the Division of Manufacturing Processes and Automation Engineering at Netaji Subhas University of Technology, New Delhi. He completed his Ph.D. from the Department of Mechanical and Industrial Engineering, Indian Institute of Technology, Roorkee. His major research areas include processing of polymer composites and green composites, machining aspects of composites, and FEM modeling of composites. He has published several research articles in journals of repute. He has also contributed many book chapters published by noted publishing houses. He is currently working as an editor of two contributed volumes on processing of polymer composites.

Dr. Kuldeep Panwar is currently working as Associate Professor and Head of Mechanical and Automation Engineering Department at Delhi Technical Campus, Greater Noida. He completed his Bachelors of Engineering from B.T.K.I.T, Dwarahat (erstwhile KEC) in 2009, and his Masters in Thermal Engineering in 2012. He completed his Ph.D. in Mechanical Engineering from G. B. Pant University of Agriculture and Technology, Pantnagar in 2016. His doctoral research area was analysis of thermal and resistance characteristics of fixed-bed regenerator with small

D/dp ratio. He has published more than 30 research articles in peer-reviewed international journals and conferences. He has one design patent to his credit. He has authored a book titled *Effect of Insolation on Efficiency of Solar Plate Collectors* and has also contributed as an editor in four conference proceedings.

Chapter 1
Dry Sliding Wear Behaviour of Aluminium Metal Matrix-Based Functionally Graded Materials Reinforced with Alumina Particles

Bhupendra Kumar, Krishna Kant Singh Mer and Lalta Prasad

Abstract In this study, a relation between reinforcement particles' quantity variation and wear behaviour along with hardness properties of the aluminium-based functionally graded material (FGM) was established. Alumina particles were reinforced in aluminium-based functionally graded metal matrix composite, and the composites were manufactured by centrifugal casting at a constant speed. The matrix of the composite was a non commercial aluminium alloy, and Al_2O_3 particles, with an average grain size of 80 μm, were used as reinforcement. The reinforcement was varied in the range of 3.0–7.5 vol.%, and all the samples were prepared as per the ASTM standards. A pin-on-disc tribometer was used to study the sliding wear behaviour of composites. During the wear test, the speed of the disc was kept constant (0.5 m/s) and load was varied (20 and 25 N) at room temperature. The disc was made of AISI 52100 steel and acted as counter-body. In centrifugally cast metal matrix composite, the accumulation of alumina particles was found towards the outer region of the casting, which also showed an increase in the microhardness in that region. Wear testing indicated comparatively less wear towards the external zone of the castings.

Keywords Metal matrix composite · Reinforcement · Wear · Friction coefficient

---

B. Kumar (✉)
Department of Mechanical Engineering, Doon Institute of Engineering & Technology, Rishikesh, Dehradun 249204, India
e-mail: bhupendra_94@yahoo.co.in

K. K. S. Mer
Institute of Technology, Gopeshwar 246424, Uttarakhand, India

L. Prasad
Department of Mechanical Engineering, National Institute of Technology—Uttarakhand, Srinagar, Garhwal 246174, India

I. Singh et al. (eds.), *Trends in Materials Engineering*,
Lecture Notes on Multidisciplinary Industrial Engineering,
https://doi.org/10.1007/978-981-13-9016-6_1

1.1 Introduction

Functionally graded materials (FGM) are newly introduced composite materials that are being researched and developed for the purpose of increasing super heat resistance of various parts such as engine parts and outer body of space crafts. In conventional super heat-resistant materials, the heat-resistant ceramic tiles such as exterior tiles of space shuttles are bonded on metal structures but the ceramic tiles tend to peel off or crack due to variation in thermal stresses, difference of thermal expansion, and other thermal properties at the contact surface between the ceramic and the metal. Most composite materials also show the same behaviour at the molecular level. Composite materials may alleviate the situation in certain circumstances by combining dissimilar materials in the bulk and not along a specific interface in order to synergize them to obtain unique combination of properties while compensating the deficiencies of one another. The FGMs can give the designer the opportunity to overcome the limitations associated with the conventional materials. The FGM can also be used as the material whose properties can be customized for given design requirements. Hence, composite materials can be used to achieve better and cheaper solutions for the established designs [1–4].

The metal matrix composites (MMCs) give the specific physical, mechanical and thermal properties, at elevated and ambient temperature as compared to conventional materials. The required properties, e.g. low density, high thermal conductivity, good fatigue responses, high abrasion and wear resistance, etc., are modified by the MMCs. Some application of MMCs is fabrication of the spacecraft, satellites, missile nose, helicopters, automobile parts, e.g. piston, connecting rod, nozzles, etc.

Centrifugal casting is considered to be one of the most useful methods that are used for processing of functionally graded materials (FGMs). It has been used extensively for FGMs which are made of aluminium matrix composites (AMCs) [5–7]. The composition gradient was obtained by using centrifugal casting which helped in distinguishing the phases with different densities [7, 8]. Denser and cleaner metals are obtained by centrifugal casting as heavier reinforcement particles are thrown to the parts of the mould from the centre of rotation and lighter reinforcement particles and impurities go to the centre. As the centrifugal force varies in radial direction, the microstructural characteristics of centrifuged FGM were investigated in radial direction [9, 10]. Earlier research has shown that heavier ceramic particle distribution in melt is influenced by various forces such as heating temperature of mould, furnace temperature of crucible and velocity of rotation of the mould. So controlling these factors result in the homogeneous metal matrix composites with higher volume fraction of reinforcement particles. Thus, casting with high density, close grain structure, good details and good mechanical properties are obtained.

The dry sliding wear of centrifugally cast homogeneous Al–SiC composites has been researched and published by many authors. Some authors concluded that the hardness behaviour of the centrifugally cast functionally graded Al–SiC composites metal matrix Al (2124)–SiC has smoother variation of SiC particles compared to Al (356)–SiC due to higher freezing range. The maximum hardness obtained at the outer periphery after heat treatment for Al (356)–SiC [8]. When the centrifugal casting of FGM was done at low centrifugal speed, the distribution of reinforcement particles presented a smooth gradient, while at higher centrifugal speed, the distribution of reinforcement particles revealed a sharp gradient [11]. The comparison of homogeneous Al–Si–Mg-20 vol.% SiC composite was done in pin-on-disc test under dry friction condition. The wear coefficient of FGM is one order less than that of homogeneous composite [12]. The mechanical and fatigue properties of AMCs were significantly increased with the increase in reinforcement ratio and decrease in reinforcement particle size. The fracture toughness of AMCs was decreased with the increase in volume fraction of Al_2O_3 [13].

In the present paper, the wear behaviour and hardness of aluminium metal matrix-based functionally graded materials reinforced with Al_2O_3 particles were investigated. Centrifugal casting was used to prepare MMCs at a constant rpm in various vol.% of Al_2O_3.

1.2 Experimental Procedures

1.2.1 Material Processing

The matrix of the composites was a non commercial Al alloy (Si = 0.7%, Cu = 0.10%, Mn = 0.30%, Mg = 0.9%), as reinforcement Al_2O_3 particles with different volumetric fractions 3.0, 4.5, 6.0 and 7.5 vol.% having 80 μm average grain size. Mg was used as a binding material.

Two steps were used to produce FGM composites. (1) stir casting and (2) centrifugal casting. In the first step, homogeneous MMCs were produced using stir-casting process, and in another step, the centrifugal casting was used to process the FGMs.

The metal was heated in the pit furnace till it gets melted. Then, the reinforcement (Al_2O_3) was added with different volumetric percentage. After that, the reinforcement was homogeneously mixed by stirrer. The mould was preheated before the metal was poured into it and the cylindrical mould was rotated with a constant rotating speed of 780 rpm. Then, the cast was cooled at room temperature, as shown in the schematic figures. The experiment was repeated with different volumetric fractions of reinforcement Al_2O_3 particles (Fig. 1.1).

The final FGM is a ring having radius of 42 mm and thickness of 15 mm. The cast ring was divided into two sections, inward and outward having a thickness of 7 mm, each for the evaluation of particle distribution in inward and outward sections of composite. Eight cylindrical specimens were prepared with dimensions as 30 mm length and 6 mm diameter.

Fig. 1.1 **a** Centrifugal casting process, **b** stir-casting process and **c** cast product

1.2.2 Sliding Wear Test

Before the sliding wear test, tungsten filament-based SEM (LEO 435VP) was used to characterize the particle distribution in the FGM samples. The samples were inspected on top surface and from inner zone to outer zone. The crystallographic characteristics of the all types of coatings were investigated with X-ray diffractometer (BRUKER AXS D8 ADVANCE, Cu kα radiation). A standard θ–2θ trace was recorded in the range of 10°–120° for complete phase identification. A Cu Kα radiation source was used ($\lambda = 1.5406$ Å), with a 40 kV accelerating voltage and a 30 mA filament current. The signal was recorded at 0.02° steps, with a dwell between 4 and 20 s at each step, according to the desired data quality. The variation of angle (2θ) was from 5° to 120°, and the scanning rate was 2° per minute.

A TR-20 LE pin-on-disc tribometer was used to perform the sliding wear test without lubrication on the pure Al alloy and FGMs. It was tested for confirmation to ASTM G99 test procedure. As counter-body a bearing steel (AISI 52100), hardness 58-62 HRc, with wear track diameter 20 mm was used. The applied normal load was $F = 20$ and 25 N, respectively, in different experiments with 0.5 m/s sliding speed. The sliding time for all the experiments was 1500 s, and all these experiments were performed at room temperature. Time was same for all specimens and for the second run of the experiment.

The TESTER is stand-alone floor model with independent mobile controller unit which facilitates study of friction and wear characteristics in sliding contact under different conditions. The principle is to press a stationary upper specimen which is cylindrical pin over a hardened rotating disc. The characteristics of material were evaluated using generated wear and online friction. Sliding occurs between the stationary specimen and a rotating hardened disc. Different variables which can be used for the test are: rotational speed, load and sliding track diameter, out of which only load was varied in our tests. The tangential frictional force and wear on specimen are continuously monitored with electronic sensor and recorded on PC as functions of time. Sensor gives the information of frictional force, wears, and speeds of part of the testing unit.

The wear coefficient (or specific wear rate) K of the samples was calculated by using Eq. (1.1) as given below:

$$K = \frac{V}{F \times X} \tag{1.1}$$

where

- V wear volume (mm^3); the wear volume was calculated by weight loss method. A Mettler electronic balance was used for weight measurements.
- F applied force (N) which is 20 and 25 N in two different runs, respectively.
- X total distance (m) which the specimen slides during the test for the time period of 1500 s.

The hardness was tested by Brinell cum Vickers hardness tester machine (HPO-250) which is suitable for testing of materials and alloys which are hard or soft, flat or round shape. Machine hardness scales are HV and BHN.

1.3 Results and Discussion

1.3.1 Microstructural Characterization

The microstructures of the reinforced Al_2O_3 sample and pure Al alloy sample are presented in Fig. 1.2. In the pure Al alloy sample, there is no reinforcement particles as shown in Fig. 1.2a and b, but in the sample with different volume percentages of Al_2O_3 reinforcement the area fraction of the postulates increases as the volume percentage of reinforcement increases as shown in Fig. 1.2d and f. On examining the inner zone and outer zone, the area fraction of postulates increases from inner to outer zone due to centrifugal force as shown in Fig. 1.2c–f. Further, we have observed that the area fraction of postulates is less in inner zone as compared to outer zone in the lower reinforcement samples (3% vol.). As the volume percentage of reinforcement increases (7.5% vol.) there is smooth variation in inner zone to outer zone due to low centrifugal speed [11].

1.3.2 Composite Characterization

X-ray diffraction was performed to determine the phase structure of the composites. Following are the X-ray diffraction patterns of centrifugally cast FGM samples:

From the analysis of the peaks shown in Fig. 1.3, it can be observed that there are two phases present in the composite; which are the Al matrix phase and the reinforcement $A1_2O_3$ phase. There are no other intermediate phases produced during fabrication. XRD analysis confirms the uniform distribution and presence of constituents in the blended powder of the specimen.

The X-ray diffraction pattern of $A1_2O_3$ (3.0, 4.5, 6.0 and 7.5 vol.%) composites along with pure aluminium was ascertained from the study. XRD patterns of pure

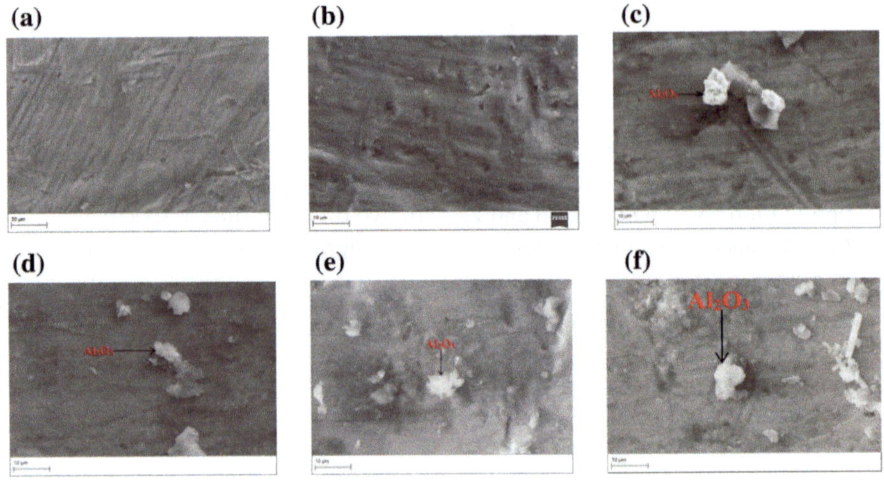

Fig. 1.2 **a** and **b** SEM micrograph for pure Al alloy; **c** and **d** inner and outer zones of 3.0% vol of Al_2O_3 reinforcement; **e** and **f** inner and outer zones of 7.5% vol of Al_2O_3 reinforcement

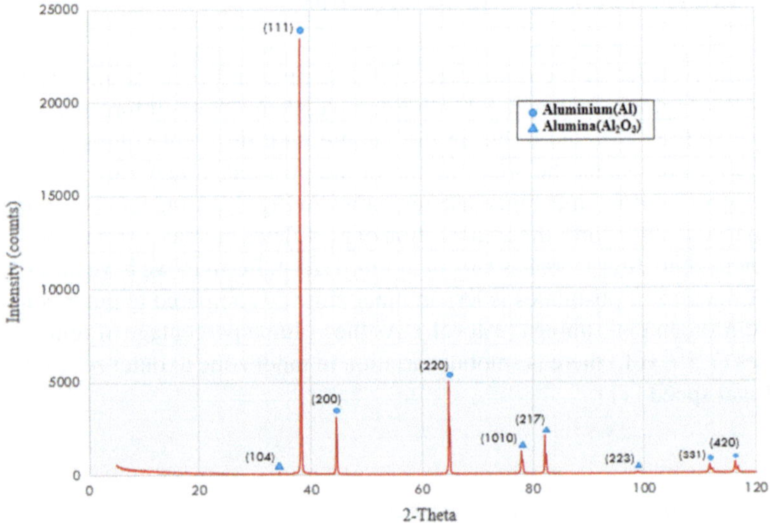

Fig. 1.3 Composite characterization of FGM

aluminium showed well-defined diffraction peaks at values of 38.43°, 44.76°, 65.02°, 111.83° and 116.36° corresponding to the reflections from the (111), (200), (220), (331) and (420) planes with face-centred cubic (FCC) structure. After adding of alumina particles into the Al matrix, new XRD peaks are observed at values of 35.2°, 78.21°, 82.33° and 97.52° corresponding to the reflections from the (104), (1010), (217) and (223) planes with rhombohedral structure. Both the XRD patterns were in accordance to diffraction JCPDS FILE NO-04-0787 and JCPDS FILE NO-81-1667.

1.3.3 *Wear Behaviour*

1.3.3.1 Wear in Micrometre with Respect to Time

We examined the wear (μm) with respect to time (s) for the 1500 s. The values of wear for pure Al alloy sample were gradually increasing in the given time intervals. The wear was increased with time, but values of wear were fluctuating at some instant of time intervals. This is due to the contact of counter-body with the protruding Al_2O_3 particles. Al alloy is a ductile metal which is used as a metal matrix, since ductile metals are prone to transfer back and forth between the sliding materials so the graph exposed positive and negative variations. The greater fluctuation has been shown in Fig. 1.4a and b inner zone samples in comparison with outer zone sample because the reinforce particles in the outer zone are greater. Obviously, the inner zone of ring is more ductile than outer zone. Here, we have provided only the graph observations of the samples at 7.5 vol.% of Al_2O_3 as the other samples also show the same behaviour in their wear–time fluctuation, and for every sample (at different vol.%), the graph of inner zone showed the greater wear–time fluctuation in comparison with outer zone.

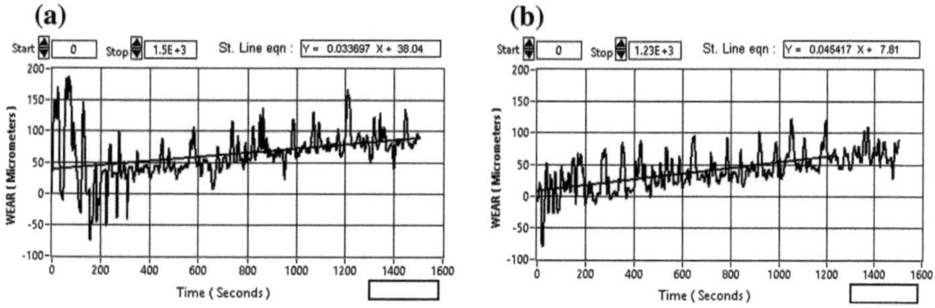

Fig. 1.4 **a** Wear in micrometres with respect to time for inner zone of ring for a sample of 7.5 vol.% (Al_2O_3). **b** Wear in micrometres with respect to time for outer zone of ring for a sample of 7.5 vol.% (Al_2O_3)

1.3.3.2 Wear Volume

Figure 1.5a shows wear volume for the 20 N run. From the figure, it is inferred that the pure Al alloy has the highest amount of wear volume in comparison with Al_2O_3 metal matrix composite, as such wear volume of sample with 3.0% vol of Al_2O_3 is less than the pure Al alloy and similar wear volume results are obtained for the increasing the volume percentage of the reinforcement as 4.5, 6.0 and 7.5%. This is due to increase in the area fraction of Al_2O_3 particles in Al matrix. Reinforced particles are usually protruding out from the metal matrix composite surface which results in increased surface roughness and also provides protection to the matrix from effective contact with counter-body [14].

On examining the inner zone and outer zone of various samples, it is observed that the wear volume is more for the inner zone sample than the outer zone sample. In outer zone, the Al_2O_3 particles have more area fraction as compared to inner zone. The centrifugal force caused the movement of heavier Al_2O_3 particles from inner zone to outer zone. The difference of wear volume in inner zone and outer zone is $\approx$0.46 mm^3 for 3.0% vol. samples. This difference decreases on increasing the volume percentages (4.5, 6.0 and 7.5%) of Al_2O_3 reinforcement, i.e. $\approx$0.15 mm^3 in 7.5% vol. samples. In this FGM cast, the average speed ($\approx$800 rpm) was taken which results in more smoother variation of the reinforcement from inner to outer zone.

To verify the first run results, we performed the second experiment under a load of 25 N as shown in Fig. 1.5b. It is shown that the wear volume is more as compared to the first run in each sample due to increased load. The wear volume differences between inner and outer zones are decreasing as the volume percentage of reinforcement is increasing as in the first run under a load of 20 N.

1.3.3.3 Wear Coefficient Variation

The tabulated values of average wear coefficient were studied in two runs. In the first run, the applied load was 20 N and a load of 25 N was applied in the second run. The wear volume was calculated by the weight loss method. The total sliding distance

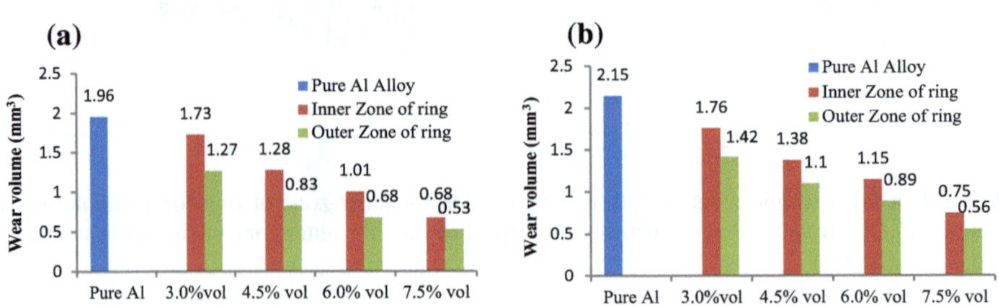

Fig. 1.5 **a** Wear volume of various samples in first run with 20 N load. **b** Wear volume of various samples in second run with 25 N load

was ≈754 m. The value of wear coefficient decreases from inner to outer zone of the ring. It is due to the presence of more reinforcement particles in the outer zone due the effects of the centrifugal force. The other reasons may be the higher density of reinforcement particles in 7.5 vol.% as compared to 3.0 vol.% of reinforcement particles. It may result in higher collision between the particles which causes the smoother variation of particles [11, 15]. The value of wear coefficient in the inner and outer zones decreases as the amount of reinforcement particles increases from 3.0 to 7.5% in volume percentage as shown in Table 1.1.

To verify the above results, the same experiments were carried out under an applied load of 25 N. In the second run, the wear volume was more as compared to the first run, but at the same time the applied load was more as compared to the first run. So the wear coefficient was varying slightly with respect to the first run as shown in Table 1.2.

1.3.4 Vickers Hardness Test

On investigating the results of Vickers hardness test for the Al alloy and the FGM samples with different vol.% of Al_2O_3 (3.0–7.5 vol.%), we observed the higher densities of reinforcement particles in the outer zone as compared to inner zone of the cast ring. As the vol.% of the reinforcement increases, the hardness of the FGM samples increases because the postulates of the reinforcement increase the density of

Table 1.1 Wear coefficient for various samples (first run with 20 N load)

S. No.	FGM ring zones	Wear coefficient (mm^3/Nm) of Al alloy (pure)[a]	Wear coefficient (mm^3/Nm) of reinforcement, Al_2O_3			
			3 (vol. %)	4.5 (vol. %)	6 (vol. %)	7 (vol. %)
1	Inner zone	0.0001302	0.000115	0.0000849	0.0000669	0.0000451
2	Outer zone	0.0001302	0.000082	0.0000550	0.0000450	0.0000351

[a]Casting was performed taking a single sample, i.e. Al alloy (pure) in first run

Table 1.2 Wear coefficient for various samples (second run with 25 N load)

S. No.	FGM ring zones	Wear coefficient (mm^3/Nm) of Al alloy (pure)[a]	Wear coefficient (mm^3/Nm) of reinforcement, Al_2O_3			
			3 (vol. %)	4.5 (vol. %)	6 (vol. %)	7 (vol. %)
1	Inner zone	0.0001139	0.0000936	0.0000736	0.00006128	0.00003993
2	Outer zone	0.0001139	0.00007533	0.00005835	0.00004736	0.0000297

[a]Casting was performed taking a single sample, i.e. Al alloy (pure) in second run

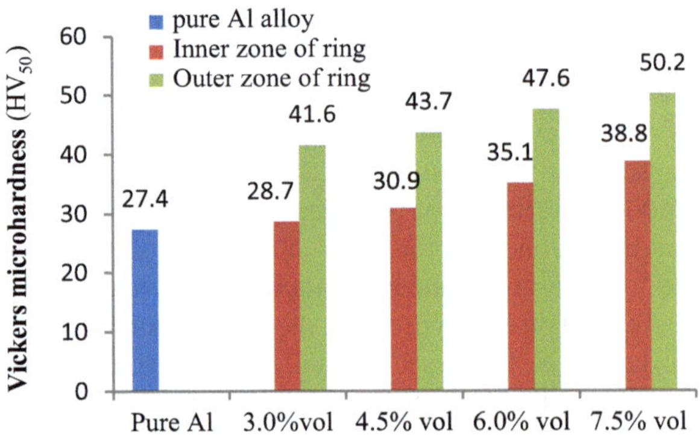

Fig. 1.6 Vickers microhardness (HV50) values of various samples

Al_2O_3 particles in the outer zone. The hardness values of the samples were increasing from the inner zone to outer zone. The main reason for the increase in the hardness is due to the presence of the composition gradient in the composite material. The gradients in the composite also play an important role and affect the properties like thermal expansion coefficient and Young's modulus as reported by many researchers [16, 17]. The result analysis shows increase in hardness from inner zone to outer zone of the samples, and as the vol.% of Al_2O_3 reinforcement particles increases, the difference between the hardness of inner to outer zone decreases as shown in Fig. 1.6.

1.4 Conclusions

In this experimental study, the mechanical and tribological properties along with microstructural performance tests of Al metal matrix composite reinforced with micro-alumina particles were investigated. The composites were fabricated via centrifugal casting route at constant speed (780 rpm). On examining the results, we see that up to 7.5 vol.% of reinforcement particle material behaves like a functionally graded material (FGM). On going towards the 7.5 vol.% of reinforcement particles, the difference between wear rate and hardness of inner and outer zones of FGM is very less. On increasing the vol.% of reinforcement, the difference between the wear rate and hardness of inner and outer zone is decreasing. Further on addition of the more vol.% of reinforcement particles beyond 7.5 vol.%, the difference between wear rate and hardness of inner and outer zones would tend to be negligible, as indicated by the wear volume graph. It means material would behave like a non-functional material. For more addition of reinforcement particles, we have to increase the speed of revolution in centrifugal casting process, because it would result in sharper penetration of particles.

References

1. Aboudi, J., Pindera, M.J., Arnold, S.M.: Higher-order theory for functionally graded materials. Compos. B **30**, 777–832 (1999)
2. Agarwal, B.D., Bansal, R.K.: Plastic analysis of fibre interactions in discontinuous fibre composites. Fibre Sci. Technol. **10**, 281–297 (1977)
3. Giannakopulos, A.E., Suresh, S., Olsson, M.: Elastoplastic analysis of thermal cycling: layered materials with compositional gradients. Acta Metall. Mater. **43**(4), 1335–1354 (1995)
4. Agarwal, B.D., Broutman, J.: Three-dimensional finite element analysis of spherical particle composites. Fibre Sci. Technol. **7**, 63–77 (1974)
5. Wang, K., Xue, H., Zou, M.: Microstructural characteristics and properties in centrifugal casting of SiCp/Zl104 composite. Trans. Nonferrous Met. Soc. China **19**, 1410–1415 (2009)
6. Duque, N.B., Melgarejo, Z.H., Suarez, M.O.: Functionally graded aluminium matrix composites produced by centrifugal casting. Mater. Charact. **55**(2), 167–171 (2005)
7. Kieback, B., Neubrand, A., Riedel, H.: Processing techniques for functionally graded materials. Mater. Sci., Eng. A **362**, 81–106 (2003)
8. Rajan, T.P.D., Pillai, R.M., Pai, B.C.: Functionally graded Al-Al3Ni in situ intermetallic composites, fabrication and microstructural characterization. J. Alloy. Compd. **453** (2008)
9. Watanabe, Y., Yamanaka, N., Fukui.: Control of composition gradient in a metal-ceramic functionally graded material manufactured by the centrifugal method. Compos. A **29**, 595–601(1998)
10. Drenchev, L., Sobczak, J. Sobczak, N.: Sedimentation Phenomenon and viscosity of water-SiC suspension under gravity Conditions-a water model study for composites synthesis. Colloids. Surf. A **197**, 203–211 (2002)
11. Vieira, A.C., Sequeira, P.D., Gomes, J.R., Rocha, L.A.: Dry sliding wear of Al alloy/SiCp functionally graded composites—influence of processing conditions. Wear **267**, 585–592 (2009)
12. Gomes, J.R., Rocha, L.A., Crnkovic, S.J., Silva, R.F., Miranda, A.S.: Friction and wear properties of functionally graded aluminum matrix composites. Mater. Sci. Forum **423–425**, 91 (2003)
13. Asif Iqbal, A.K.M., Nuruzzaman, D.M.: Effect of the reinforcement on the mechanical properties of aluminium matrix composite. A review, Int. J. Appl. Eng. Res. **11**, 10408–10413 (2016)
14. Mondal, D.P., Das, S., Rao, R.N., Singh, M.: Effect of SiC addition and running-in wear on the sliding wear behaviour of Al–Zn–Mg aluminium alloy. Mater. Sci. Eng. **402**, 307–319 (2005)
15. Gao, J.W., Wang, C.Y.: Modeling the solidification of functionally graded materials by centrifugal casting. Mater. Sci. Eng. **292**, 207–215 (2000)
16. Watanabe, Y., Fukui, Y.: Microstructures and mechanical properties of functionally graded materials fabricated by a centrifugal method. Rec. Res. Develop. Metall. Mater. Sci. **4**, 51–93 (2000)
17. Humberto Melgarejo, Z., Marcelo Suarez, O., Sridharan, K.: Wear resistance of a functionally-graded aluminium matrixcomposite. Scripta Mater. **55**, 95–98 (2006)

Chapter 2
Fracture Toughness of Symmetric and Asymmetric Layup GFRP Laminates by Experimental and Numerical Methods

Ruchir Shrivastava and K. K. Singh

Abstract FRPs are widely used composite materials, and they offer great advantages of reduction in weight compared to conventional materials. Delamination is seen as one of the biggest causes of its failure. The work herein examines fracture behavior pattern of GFRP laminates, made by reinforcing E-glass plain weave (PW) fiber and epoxy as matrix material. The Plain weaved fiber finds large acceptability in composite industries due to their flexibility, better handling capacity during manufacturing, and better damage tolerance. The laminate was configured in symmetric and asymmetric conditions and made by hand layup assisted by vacuum bagging technique. The laminates configured are quasi-isotropic with zero bending-extension stiffness coupling matrix. Mode-I test was conducted following ASTM D5528 standard in a computerized UTM in displacement control mode. The numerical analysis was done using cohesive zone modeling. The results showed better fracture toughness for asymmetric layup which was due to extensive fiber bridging in these kinds of layups. The numerical results were able to mimic the delamination behavior, and close similarity is obtained in force–displacement curve.

Keywords FRP · Delamination · Fracture toughness

Abbreviations

CF	Carbon fiber
CNT	Carbon nanotube
CSCNT	Cup-stacked carbon nanotube
DWCNT	Double-wall carbon nanotube
EP	Epoxy

R. Shrivastava (✉) · K. K. Singh
Department of Mechanical Engineering, Indian Institute of Technology (Indian School of Mines), Dhanbad 826004, Jharkhand, India
e-mail: ruchirpanduka@gmail.com

I. Singh et al. (eds.), *Trends in Materials Engineering*,
Lecture Notes on Multidisciplinary Industrial Engineering,
https://doi.org/10.1007/978-981-13-9016-6_2

HNT Halloysite nanotube
MWCNT Multiwall carbon nanotube

2.1 Introduction

In recent decades, FRP composites became popular in aeronautics, marine and structural applications. They offer advantage of high strength-to-weight ratio, high stiffness and tailorable properties. In view of making these composites more reliable, a lot of research is focused in these days to better characterize its material property, in understanding of its failure mechanisms and improvement of life performance and its durability. There have been plenty of works focusing on these problems with experimental and numerical techniques. There have also been continuous efforts to improve strength properties of these composites which included efforts made by nanoparticle inclusions such as nanosilica, carbon black, fumed silica, fullerene, DWCNT, MWCNT, HNT, CSCNT nanoclay and organoclay. The property analysis particularly focusing on mechanical properties depicts enhancements in properties such as fracture toughness, energy absorption capacity, tensile strength, compressive strength and shear strength. The achieved enhancement depends upon a lot of parameters such as quantity and type of nanoparticle, dispersion achieved, fiber and matrix strength and their interfacial bonding characteristics. The fiber–matrix damage mechanism plays a significant role in this aspect, and the specimen failure pattern after the first fiber breakage is vital for understanding the complete failure mechanism.

Fracture is one of the complex failure modes in composites and is comparatively a new area of research. The anisotropic nature of composites further extends the complexity in crack propagation pattern. In real life applications, the crack formation in a composite product may occur due to manufacturing defects, mishandling during life period, tool drop, etc. The sudden fracture pattern raises concern for safety reasons, and to avoid these scenarios material of very high strength capacity is used. This increases the weight of the component, and hence complete utilization of composite properties is restricted.

Based on the loading and crack propagation direction, fracture can be categorized into three types, opening mode (Mode-I), shearing mode (Mode-II) and tearing mode (Mode-III). The fracture toughness of FRP composites is measured by its strain energy release rate [1]. It is a matrix-based property, and hence its performance can be improved by matrix toughening via nanoparticle addition [2]. Fiber surface treatment [3], alteration of fiber volume fraction [4] and improvement in manufacturing methods [5] are some of the other possible methods. The stacking sequence of a laminate is an important parameter, and its variation can impact material strength to a great extent [6, 7].

2.2 Stacking Sequence Variation

Azzam and Li [8] studied the effect of stacking sequence on bending strength of carbon/epoxy laminates using three-point bend test (ASTM D7264/D7264M-07). The laminate preparation was done using compression molding technique with heat application. Quasi-isotropic and unbalance stacking sequences were manufactured by placing laminates in $[0°/90°/-45°/45°]_{2s}$ and $[45°/45°/90°/0°]_{2s}$ stacking sequence. Brittle failure was seen for quasi-isotropic specimen, while fiber failure, delamination and debonding were reported for unbalanced specimen. Samborski [9] studied the effect of boundary condition on Mode-I distribution along the width of the specimen. Bending testing (BT) coupling was chosen and compared with bending-extension (BE) coupling. Fiber orientation angle (α) was chosen as 0°, 30°, 45°, 60° and 90°. The results obtained through Abaqus© software using VCCT showed symmetric and flat distribution of strain energy release rate (SERR) along the width of UD and BE ($\alpha = 90°$) specimen. Results for other specimens concluded that error in G_{IC} calculation may occur by ASTM D5528 standard due to the non-symmetric distribution. Mixing of mode failure may also result, which can give inaccurate G_{IC} values. Huddhar et al. [10] studied the effect of varying pre-crack length on interlaminar fracture toughness of glass/epoxy laminated composites. The calculation of Mode-I (ASTM D5528-94a) fracture toughness was done using beam theory and compliance calibration while Mode-II (JIS standard) fracture toughness by beam analysis method. Results showed increment in Mode-I ILFT by 11.12 and 11.74% with pre-crack length increase from 20 to 30 mm (50% enhancement) using modified beam theory and compliance method, respectively, while Mode-II ILFT found to be decreased by 62.26% with the same variation of crack length. The load-carrying capacity was found better for short pre-crack length. Moreno et al. [11] conducted three-point bend test for CF/EP composites in view of analyzing its tensile and compressive behavior in a flexural test. Two unidirectional $[0°]_5$ and $[0°]_8$ and three symmetric cross-ply $[90°_2, 0°_2]_s$, $[0°_2,90°_2]_s$ and $[0°_3, 90°_2]_s$ laminates were made with an average thickness of 0.25 mm. Modifications in the cross section were done, and static moment was zero. The most deformation prone zone is the distant layer from neutral layer, due to bending, whereas the interlaminar failure may grow along neutral layer. The results for unidirectional five plies and eight plies showed flexural failure and shear failure, respectively, indicating effect of thickness on obtained results. For cross-ply laminates, one and two, the stacking sequence of the laminates played a significant role and bending moment in the upper layers was found first, ahead of which complete failure is processed. For third cross-ply, start of crack was found at the neutral axis. Rehan [12] studied variation of stacking sequence at crack interface and its effect in Mode-I ILFT. The laminates prepared and cured through hot compression are quasi-isotropic and quasi-homogeneous in nature to accomplish removal of other parameters which can affect fracture behavior. (90°/45°//45°/90°) and (−45°/45°//45°/−45°) showed higher fiber bridging phenomenon and hence resulted in better fracture toughness values. interface and subsequent layers are found to be affecting the fracture behavior. It could be seen that the material performance was greatly affected by the chosen stacking sequence. The literature survey indicated less of work particularly for PW fabric, and hence in this presented work, interlaminar fracture toughness determination of these fabrics is reported using glass fiber.

2.3 Layup Configuration

Symmetric and asymmetric, two kinds of quasi-isotropic layups were chosen for this study (Fig. 2.1). The forces and moments acting in a laminate are analyzed by classical laminate theory [13]. [*A*], [*B*] and [*D*] matrices can also be calculated by U20MM [14]. The selected layups have zero bending stiffness.

$$\left\{\frac{N}{M}\right\} = \left\{\frac{A}{B}\middle|\frac{B}{D}\right\}\left\{\frac{\varepsilon^0}{k}\right\} \tag{2.1}$$

where

N Force.
M Moment.
[*A*] Extension stiffness matrix.
[*B*] Bending-extension coupling stiffness matrix.
[*D*] Bending stiffness matrix.

The resulting layups are:

1. [(0°, 90°)/(+45°, −45°)/(+45°, −45°)/(0°, 90°)//(+45°, −45°)/(0°, 90°)/(0°, 90°)/(+45°, −45°)] for asymmetric laminate.
2. [(0°, 90°)/(+45°, −45°)/(+45°, −45°)/(0°, 90°)//(0°, 90°)/(+45°, −45°)/(+45°, −45°)/(0°, 90°)] for symmetric laminate.

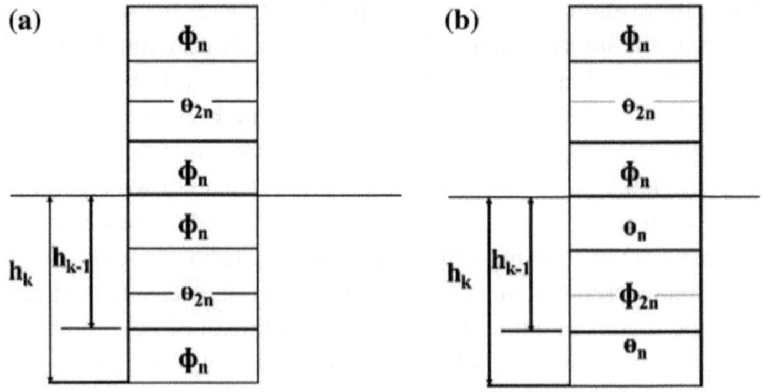

Fig. 2.1 **a** Symmetric and **b** asymmetric layup configuration [15]

2.4 Experimental Setup

The Mode-I ILFT experiment (Fig. 2.2a) was conducted using DCB test specimen under displacement control on a 50 kN Hounsfield at a rate of 1 mm/min. All the experiments were conducted under normal laboratory conditions. Laminates were prepared using hand layup assisted by vacuum bagging technique (Fig. 2.2b), and pre-crack of length 63 mm was generated using a Teflon film of 13-μm thickness. Eight layered laminates were manufactured, and specimen sizes were 140 × 25 × 4 mm^3. The observation of crack length was made by USB digital traveling microscope. The data reduction was done by compliance calibration and modified beam theory (Eq. 2.2).

$$G_I = \frac{3P\delta}{2b(a + \lfloor\Delta\rfloor)} \tag{2.2}$$

where

Δ—correction factor for crack tip rotation, *P*—load, *a*—delamination length, *w*—specimen width, *δ*—displacement.

2.5 Numerical Background

2.5.1 Description of Numerical Model for DCB Specimen

The finite element model was developed in Abaqus 6.14.1 software. Modeling of the GFRP specimen was done using 3D built-in continuum shell elements (CS8R) and meshed using continuum shell element of eight nodes and reduced integration (CS8R), and the cohesive zone was formulated using three-dimensional cohesive

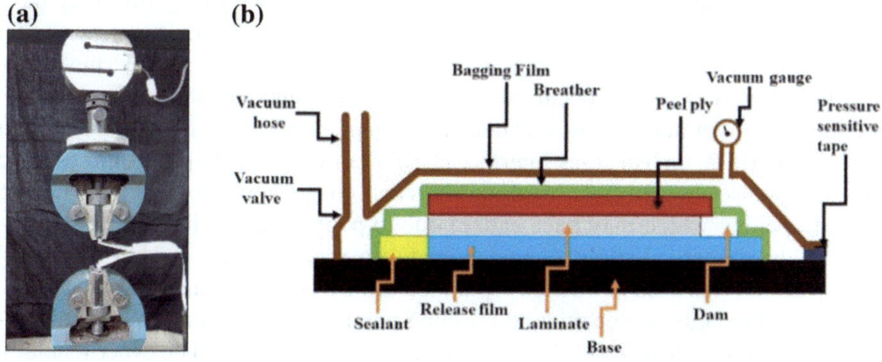

Fig. 2.2 **a** Experimental setup and **b** hand layup technique assisted by vacuum bagging

element COH3D8. Fiber and matrix failure was predicted using Hashin damage criterion. Element deletion was used to remove damaged elements in cohesive layer during meshing.

2.5.2 Cohesive Zone Modeling

Virtual crack closure technique is a commonly used FEM technique for analysis of strain energy release rate (SERR). The first few assumptions of the numerical study were made by Rybicki and Kanninem [16], and mathematical derivation was given by Raju [17]. In this study, this technique is used in Abaqus© software.

2.6 Result and Discussion

The load–displacement graph for both the layups is given (Fig. 2.3a). It was seen that asymmetric layups show higher maximum load-carrying capacity than symmetric layups. The typical load–displacement behavior shows initial linearity which then shows sudden drop of load due to crack initiation. The onset values of ILFT for symmetric and asymmetric layups were 783.53 and 537.0264 J/m^2, respectively (Fig. 2.4). The obtained result depicts that initial values are independent of fiber bridging phenomenon. The propagation values for a crack length of 83 mm were 681.04 and 1473.29 J/m^2, and hence relative increase of 116.33% was observed. The interlaminar fracture toughness values were calculated by data reduction methods of compliance calibration and modified beam theory. Due to their almost identical results, only modified beam theory results are reported. The interlaminar fracture toughness is very sensitive to crack length in the specimen, and hence care has been taken in keeping constant crack length in all the specimens. The propagation of crack was resisted by fiber pullout, its breakage and fiber bridging phenomenon. The load–displacement graph analysis also indicates that crack propagation may have followed stick-slip pattern. The extent of fiber bridging was monitored by digital microscope, and it was found more in asymmetric specimen (Fig. 2.5). This difference in behavior may also occur because of the difference of fiber layup in crack interface. The symmetric layup had delamination plane at $\phi_n//\phi_n$ interface, although the asymmetric layup had crack interface at $\theta_n//\phi_n$ orientation [18]. The microscopic observation also showed fiber breaking and crack jumping phenomenon during the Mode-I opening which may have caused due to poor resin distribution in the laminate. In the numerical study, element size taken was 0.25 and quadratic nominal stress criterion was used in traction separation law by Quads damage command. The numerical simulation is sensitive to meshing parameters, and continuum shell elements are used by which 7000 elements are generated (Fig. 2.3b). The experimentally found mechanical properties (Table 2.1) and fracture energies are used in the numerical study. The load–displacement obtained by experimental and numerical methods

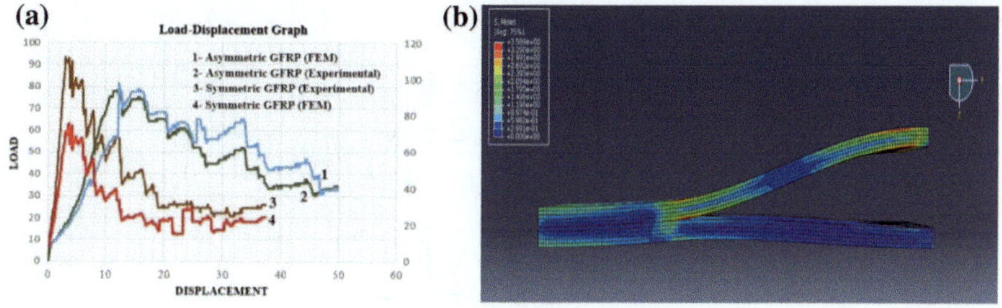

Fig. 2.3 **a** Load-displacement graph and **b** von Mises stress for asymmetric layup configuration

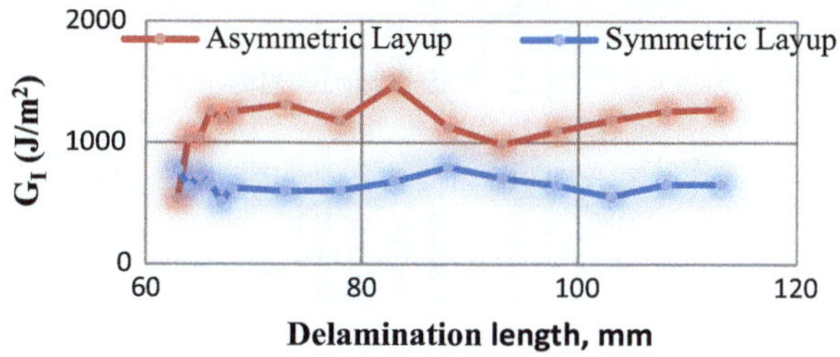

Fig. 2.4 R-curve of symmetric and asymmetric layup

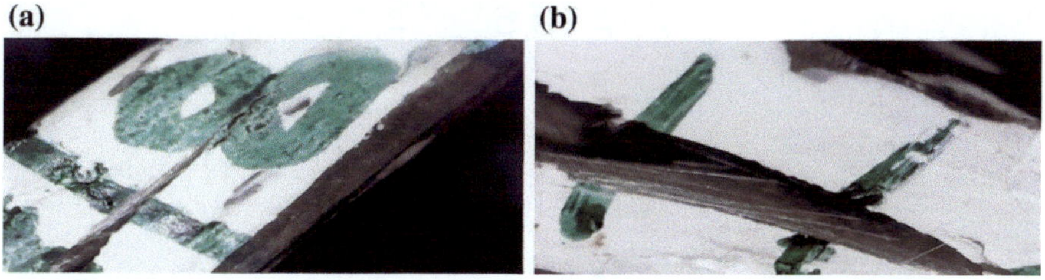

Fig. 2.5 Phenomenon of fiber bridging and fiber pullout for **a** symmetric and **b** asymmetric layup during delamination of GFRP specimen

was closely matching, although a difference in maximum load-carrying capacity of the order of 35.24 and 33.33% is reported for asymmetric and symmetric layup configuration, respectively, before crack initiation phenomenon. The assumptions made in linear elastic fracture mechanics (LEFM) approach cause the difference.

2.7 Conclusion

The laminate performance is considerably affected by the chosen laminate sequence. In this work, performance of symmetric and asymmetric GFRP laminate was examined for Mode-I ILFT test. The performance was first evaluated experimentally in a

Table 2.1 Material properties of GFRP laminates used in Abaqus 6.14.1

Longitudinal tensile strength (X_T) (MPa)	Longitudinal compressive strength (X_C) (MPa)	Transverse tensile strength (Y_T) (MPa)	Transverse compressive strength (Y_C) (MPa)	Longitudinal shear strength (S_L) (MPa)	Transverse shear strength (S_T) (MPa)	Longitudinal tensile fracture energy (G_{XT}) (N/mm)	Longitudinal compressive fracture energy (G_{XC}) (N/mm)	Transverse tensile fracture energy (G_{YT}) (N/mm)	Transverse compressive fracture energy (G_{YC}) (N/mm)
800	510	85	330	60	160	30	20.2	5	5

computerized UTM under displacement control mode, and then data reduction was done using modified beam theory. The experimental work exhibits better fracture toughness for asymmetric layups. The difference arose due to higher extent of fiber bridging in these kinds of laminates. The placement of Teflon film at mid-plane in $\phi_n//\phi_n$ and $\theta_n//\phi_n$ interfaces also developed difference in crack resistance pattern and was higher for $\theta_n//\phi_n$ interfaces. The numerical analysis further proves our hypothesis, and their results obtained for load–displacement were similar to experimental results.

References

1. Kumar, P.: Elements of fracture mechanics. Tata McGraw Hill, New Delhi, India (2009)
2. Khan, S.U., Kim, J.K.: Impact and delamination failure of multiscale carbon nanotube-fiber reinforced polymer composites: a review. Int. J. Aeronaut. Space Sci. **12**, 115–133 (2011). https://doi.org/10.5139/IJASS.2011.12.2.115
3. Rikards, R., Buchhoiz, F.G., Bledzki, A.K., et al.: Mode-I, Mode-II and Mixed-Mode I/II interlaminar fracture toughness of GFRP influenced by fiber surface treatment. Mech. Compos. Mater. **5**(32), 439–462 (1996). https://doi.org/10.1007/BF02313863
4. Adrian, F.G., Robinson, P., Pinho, S.: Effect of variation in fibre volume fraction on modes I and II delamination behaviour of 5HS woven composites manufactured by RTM. Compos. Sci. Technol. **14**(69), 2368–2375 (2009). https://doi.org/10.1016/j.compscitech.2009.02.008
5. Amirkhosravi, M., Pishvar, M., Altan, M.C.: Improving laminate quality in wet lay-up/vacuum bag processes by magnet assisted composite manufacturing (MACM). Compos. A: Appl. Sci. Manuf. **98**, 227–237 (2017). https://doi.org/10.1016/j.compositesa.2017.03.032
6. Singh, K.K., Singh, N.K., Jha, R.: Analysis of symmetric and asymmetric glass fiber reinforced plastic laminates subjected to low velocity impact. J. Compos. Mater. **14**(50) (2016). https://doi.org/10.1177/0021998315596594
7. Daricik, F., Aslan, Z.: Characterization of delamination crack in multidirectional e-glass/epoxy composite under Mode I loading. Eur. Mech. Sci. **1**, 117–128 (2017). https://doi.org/10.26701/ems.341788
8. Azzam, A., Li, W.: An experimental investigation on the three-point bending behavior of composite laminate. IOP Conf. Series: Mater. Sci. Eng. **62** (2014). https://doi.org/10.1088/1757-899x/62/1/012016
9. Samborski, S.: Effect of boundary conditions on the distribution of mode I fracture toughness along delamination front in CFRP laminates with general fiber orientation. Mech. Mech. Eng. **20**, 121–127 (2016)
10. Huddhar, A., Desai, A.: Sharanaprabhu, C.M., Kudari, S.K., Gouda, P.S.S.: Studies on effect of pre-crack length variation on inter-laminar fracture toughness of a glass epoxy laminated composite. IOP Conf. Series: Mater. Sci. Eng. **149** (2016). https://doi.org/10.1088/1757-899x/149/1/012161
11. Moreno, M.C.S., Gutierrez, A.R., Vicente, J.L.M.: Flexural testing on carbon fibre laminates taking into account their different behaviour under tension and compression. IOP Conf. Series: Mater. Sci. Eng. **139**(1) (2016). https://doi.org/10.1088/1757-899x/139/1/012047
12. Rehan, M.S.M., Rousseau, J., Gong, X.J., Guillaumat, L., Ali, J.S.M.: Effects of fiber orientation of adjacent plies on the mode I crack propagation in a carbon-epoxy laminates. Procedia Eng. **10**, 3179 (2011). https://doi.org/10.1016/j.proeng.2011.04.525
13. Jones, R.M.: Mechanics of composite materials, 2nd edn. Taylor & Francis, New York, NY (2009)
14. The University of Twente micromechanics modeller. Retrieved January 21, from http://www.opm.ctw.utwente.nl/research/pt/tools/index.html (2005)

15. Singh, K.K., Singh, R., Chandel, P., Kumar, P.: An asymmetric FRP laminate with a circular precrack to determine impact-induced damage. Polym. Compos. **29**, 1378–1383 (2008). https://doi.org/10.1002/pc.20422
16. Rybicki, E.F., Kanninem, M.F.: A finite element calculation of stress intensity factors by a modified crack closure integral. Eng. Fract. Mech. **9**, 931–938 (1977). https://doi.org/10.1016/0013-7944(77)90013-3
17. Raju, I.S.: Calculation of strain energy release rates with higher order and singular finite elements. Eng. Fract. Mech. **28**(3), 251–274 (1987). https://doi.org/10.1016/0013-7944(87)90220-7
18. Pereira, A.B., de Morais, A.B.: Mode I interlaminar fracture of carbon/epoxy multidirectional laminates. Compos. Sci. Technol. **13–14**(64), 2261–2270 (2004). https://doi.org/10.1016/j.compscitech.2004.03.001

Chapter 3
Optimization-Related Studies of EDMed Aluminum Metal Matrix Composites: A State-of-the-Art Review

Rama Nand Yadav, Rajesh Kumar Porwal and J. Ramkumar

Abstract Electrical discharge machining is a flattering non-conventional material removal process which uses energy instead of tool to remove materials. This process has capability to machine virtually any electrical conductive and extremely hard metallic materials such as metal matrix composites, aluminum metal matrix composites, duplex steel, nickel, and titanium alloys. These materials are difficult to machine by conventional process because of high cost of tool, rapid tool wear, low materials removal rate which leads to formation of build-up edges, poor surface finish and burr formations. In order to machine these materials by electrical discharge machining, machining performance will be appropriate. Many researchers have suggested a lot of appropriate ways to improved machining of these materials by analyzing the process parameters of electrical discharge machining. These studies show that machining of these materials can be improved by proper selection of process parameters. This article shows the review of experimental and theoretical studies carried out so far in the field of electrical discharge machining. This article also explored the optimization techniques used to determine process parameters for EDMed aluminum metal matrix composites.

Keywords Electrical discharge machining (EDM) · Aluminum metal matrix composite (AMC) · Optimization technique

3.1 Introduction

EDM is broadly used non-conventional machining mechanism adapted for material removal of extremely hard metallic materials by the initiation of potential difference between conductive tool material and workpiece in the existence of appropriate

R. N. Yadav · R. K. Porwal (✉)
Faculty of Mechanical Engineering, Shri Ramswaroop Memorial University, Lucknow, India
e-mail: porwal.me@srmu.ac.in

J. Ramkumar
Department of Mechanical Engineering, Indian Institute of Technology Kanpur, Kanpur, India

I. Singh et al. (eds.), *Trends in Materials Engineering*,
Lecture Notes on Multidisciplinary Industrial Engineering,
https://doi.org/10.1007/978-981-13-9016-6_3

dielectric medium. Joseph Priestly developed an erosive method using electrical discharge in 1770s, and in 1943, N. I. Lazarenko scientist from Russian developed that a succession of sparks generates between two electrically conductive materials plunged into dielectric medium without any direct contact between them [1].

EDM is a thermal-based machining process in which electrical energy is converted to thermal energy through a bombarding of distinct current materializing between conductive tool and workpiece dipped into a dielectric, in the presence of significant potential difference throughout electrodes. There is a small gap between electrode and workpiece to provide electrical resistance between the gaps. Due to this small amount of materials melts and vaporized from both electrode and workpiece material at the point of spark contact and by products of vaporization resulting a rapidly expansion of bubble. When electric pulse is terminated, heating and spark are paused; this provides vapor bubble collapsed, and the circulating dielectric fluid flushes out microscopic debris from both the electrodes.

A brief description about material removal mechanism sequences is presented in Fig. 3.1. Figure 3.1a shows the development of strong electric field, when voltage is applied between both electrodes. When voltage reaches its peak point, insulating characteristics of dielectric medium decrease along with the strongest part of the field. As the number of ionic particles increases, current is established and discharge channel is formed as shown in Fig. 3.1b. The vaporization of materials and formation of plasma channel is presented in Fig. 3.1c. After this bubble formation takes place as shown in Fig. 3.1d and when pulse voltage ceases, plasma channel collapses and

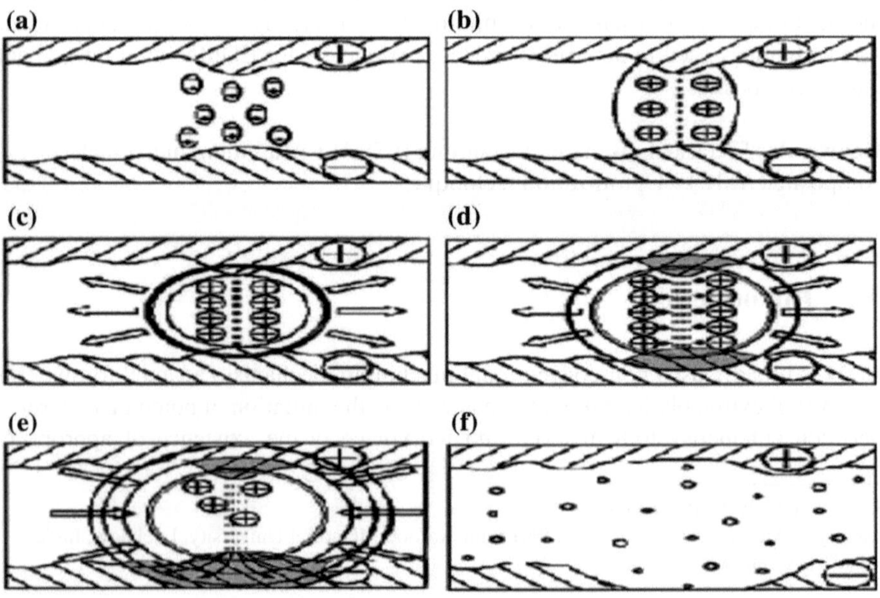

Fig. 3.1 Graphical representation of EDM mechanism [2]

molten cavities exploded into dielectric fluid as represented in Fig. 3.1e. At the last, surface cool instantaneously where vaporized and melted materials are flushed away by dielectric fluid [2].

3.2 Parameters of Electrical Discharge Machining

The input parameters such as discharge voltage, peak current, electrode gap, polarity, pulse interval (T_{off}), and pulse duration (T_{on}) while performance parameters like MRR, TWR, and surface roughness (SR) are used for EDM. Other influences of machining parameters are flushing condition, electrode, and workpiece rotation.

The amount of power used in EDM represents as peak current, and it is expressed as an increased level of current toward its preset value for every single pulse-on time. The consumed amount of electrical current is supervised on the basis of machined surface area. Higher strength current is used in roughing operations and in cavities or broad surface areas [3]. The time period, on which electrical discharge occurs, is known as pulse-on time or pulse duration. In this time, potential is applied for the machining between the tool and workpiece. On the other case, the time interval between two pulses is referred to as pulse-off time. It is the time that allowed molten metal to solidified and flushed out from arc gap portion [4]. The portion between tool and workpiece in EDM process is shown as arc gap which is basically utilized for the purpose of spark generation and controlled by EDM servo system [5]. EDM process workpiece shows either positive or negative polarity. In the positive scenario, workpiece will be on the positive terminal portion and tool on negative on the other side, negative polarity behave inverse to positive one [6]. Workpiece rotation technique is used for the proper dielectric fluid circulation and temperature distribution while electrode rotation is performed for exceed flushing action and spark [7, 8]. The removal of material from machining zone per unit time represents as MRR while TWR is wear rate of tool per unit time. The parameters like current, supply voltage, and pulse-on time are directly influenced the MRR and TWR both while pulse-off-time case is inverse.

3.3 Review on Experimental and Theoretical Studies of EDM

This section shows the experimental and theoretical studies of various EDM parameters such as T_{on}, T_{off}, dielectric medium, electrode, additive powder for different materials under various circumstances during machining from various researches.

Prabhu et al. conducted an experiment on Al-TiB_2 using EDM for machining parameters such as T_{on}, T_{off}, pulse current vibration of tool and flushing pressure. They found that discharge current influenced the MRR and TWR while higher value of current provided higher MRR, on the other hand, increased value of T_{on} provided slightly increase in TWR. They also found that white layer and induced stress both got increased with increased value of T_{on} while there are a high deviation in white layer thickness for increased pulse current [9]. Kathiresan et al. explored an experiment on AMC prepared by the combination of vortex method and pressure die-casting approach. They used copper as tool electrode for machining using T_{on}, T_{off}, voltage, and peak current as input parameters. They found that MRR and SR got improved with an increase in peak current; on the other hand, MRR got decreased with the increase in vol.% of SiC. Further, they also revealed that SR of machined surface got improved with an increase in %wt of SiC [10]. Nanimina et al. worked on AMC material to evaluate the EDM machining parameters such as peak current, T_{on}, and T_{off} for performance characteristics. They found that MRR got increased rapidly for higher T_{on} and peak current in case of Al6061 rather than AMC material while it decreases with increasing T_{off}. They also revealed that low peak current and low T_{on} provide low tool wear as compared to T_{off} [11].

Mohan et al. explored an experiment to investigate the effect of electric discharge current, T_{on}, concentration %, tool hole, and electrode rotation speed on machining properties during EDM of metal matrix composite material using a tube shape electrode. They found that rotating blind tube electrode with drilling effects provided more MRR comparing to solid electrode. They also found that the higher value of SiC% was resulted in reduced MRR, SR, and higher tool wear. Their other finding is as the rotational speed of tube electrode got higher it provided higher MRR, TWR, and better surface quality. They also conclude that MRR and tool wear rate were improved for the injection flushing method on comparing with side flushing [12]. An experiment was conducted on the Al metal matrix composite reinforced with SiC particles using an EDM process by Islam et al. They used solid and hollow copper tool electrodes and kerosene as dielectric medium for their study. The process parameters were considered as discharge current, arc gap, and electrode shape. The value of input voltage, pulse frequency, and T_{on} was constant during experiments. They observed that with hollow tool electrode, improved MRR was found at 6 A discharge current with 10-μm spark gap and with solid electrode improved MRR/TWR ratio was 21.36 at 3 A discharge current and 10-μm spark gap. They also confirm that small spark gap with solid electrode provides minimum surface roughness while the discharge current has lesser effect on workpiece surface [13].

3.4 Review on Optimization Studies of EDM

Optimization plays an important role for improving performance parameters of EDM. The appropriate value of process parameters which results desired and maximum output can be determined and identified by optimization techniques. This section deals with various optimization technique uses to explore the input parameters of EDM process [14].

Al-SiC12% metal matrix composite parametric optimization was carried out by Bhuyan et al. using EDM process. They investigate the outcome of input parameters such as T_{on}, peak current, and flushing pressure for SR, TWR, and material removal rate. They used central composite design (CCD) method under different combination of process parameters during the experiments. They used RSM method for modeling and ANOVA to check the significance of the developed model. They conducted a validation test to compare the predicted and experimental value which find the effectiveness of developed model and found the predicted error of MRR, TWR, and Ra which are within the acceptable limit. They observed that MRR and SR increased with increasing value of T_{on}, and flushing pressure while TWR decreased with them [15]. An experiment on multi-objective optimization of input parameters of Al/Si/10Mg/9 wt%, Al_2O_3/3 wt% graphite in EDM for obtaining minimum SR, minimum TWR, and maximum MRR was performed by Radhika et al. Peak current, flushing, and T_{on} were selected as input parameters for EDM. The experiment was conducted on different operating levels of process parameters according to Taguchi method and GRA which was used for the multi-objective optimization. ANOVA technique was used to conclude the significant contributing machining parameters. They observed that the most significant parameter is peak current with 61.36% contribution. The obtained optimal level of input parameters such as 20 A peak current, 100 kPa flushing pressure, and 190 μs pulse duration was found to lead to good surface finish, reduced TWR, and improved MRR in the EDM of aluminum hybrid composites [16]. Prasanna et al. used principal component analysis (PCA) for optimization of performance parameters like relative wear ratio, MRR, TWR, and SR of AA7075-SiC metal matrix composite material using EDM process. Copper material electrode and kerosene as dielectric medium were used in the experiment. The experiments were performed using various values of peak current, T_{on}, T_{off}, and voltage. They observed that by improving the value of current increased, the MRR and TWR increased with lower surface finish is shown in Figs. 3.2 and 3.3. Optimum parameters were found as current 12 A, T_{on} 15 μs, T_{off} 1 μs, and discharge voltage 35 V [17].

Anand et al. worked for WEDM on Al/$ZrSiO_4$ MMC to determine the optimal value of cutting rate using T_{on}, T_{off}, servo voltage, and peak current as process parameters. They used ANOVA and F-test for process parameters. The optimized value for maximum cutting rate was as T_{on} 120 μs, T_{off} 51 μs, peak current 169 A, and servo voltage 54 V. They found that cutting rate increases as T_{on} and pulse current increases, whereas for the increased value of T_{off} and servo voltage, cutting rate decreases. They also found large-size craters and cracks on the machined surface, as T_{on} was increased to high level and T_{off} was kept on low level [18]. The optimal process parameters of AMC was carried out by Senthil et al. using TOPSIS method considering discharge current, T_{on}, and T_{off} as input parameters. They used situ casting method for workpiece preparation and L_{18} OA approach for the input parameter optimization. They optimized performance characteristics as 0.1534 g/min MRR, 0.00034 g/min TWR, and 4.49 μm SR considering input parameters such as discharge current 35 A, T_{on} 99 μs, and T_{off} 6 μs [19].

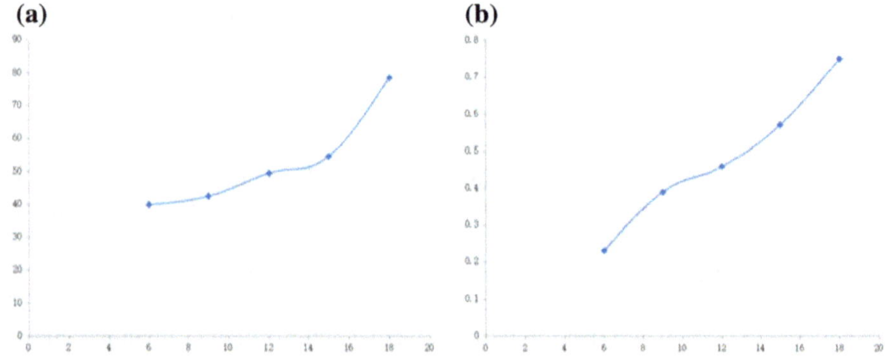

Fig. 3.2 **a** Variation of material removal rate along current and **b** variation of tool wear rate along current [17]

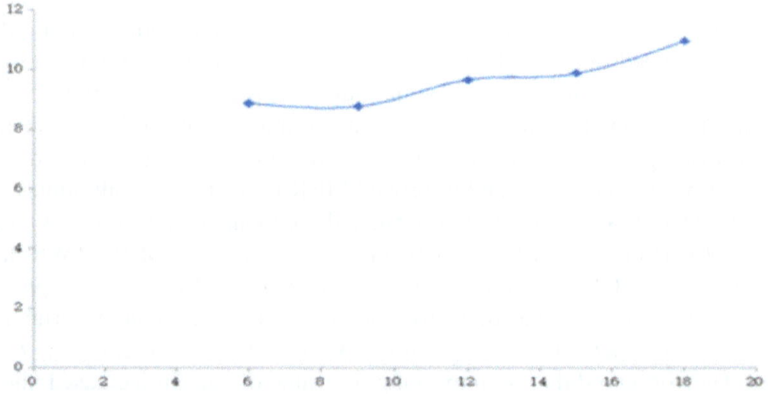

Fig. 3.3 SR variation along current [17]

Wang et al. worked to optimize the blind-hole drilling of AMC material using rotary EDM. They used Taguchi approach for the machining performance using input parameters such as polarity, ton, tool rotational speed, flushing, peak current, supply voltage, and number of eccentric through hole in electrode. The optimized value for blind-hole drilling conditions was eccentric through hole in electrode 1 hole, flushing pressure 0.6 kg/cm^2, electrode rotating speed 1000 rpm, and working time 14 min. They confirmed that blind-hole drilling EDM using rotational eccentric hole electrode provided higher MRR, and this technique is more appropriate than other machining processes [20]. The performance optimization was carried out on AMC material by Ming et al. using multi-regression modeling and two types of neural network such as backpropagation (BPNN) and radial basis. They used cutting parameters such as T_{on}, T_{off}, peak current, and servo voltage for MMR and SR. They observed that BPNN delivered best-predicted error for MRR and SR. They found that their proposed models can also show the effect of process parameters on the performance and build an intelligent optimization system with graphical user interface which was based on soft computing, GA, and desirability function. They also suggested that the optimized cutting parameters under the desired SR can obtain through their developed optimization system [21]. Kansal et al. used RSM to optimum the process parameters for PMEDMed AMC. Input parameters such as additive concentration, peak current, and T_{on} were taken for their experiment to investigate the MRR and SR. Their ANOVA outcomes show that the concentration and peak current are the most effective parameters for output characteristics. They observed maximum MRR was achieved at an additive concentration of 3 g/l. Furthermore, they also found minimum SR at a concentration between 2 and 3 g/l and at 1 A peak current [22]. Udaya et al. conducted an experiment on aluminum cast alloy reinforced with Boron carbide, to optimize the process parameters using wire EDM in which 0.25-mm-diameter Brass wire was used as the electrode. They used Taguchi technique to analyze optimal machining process parameters for minimum SR and maximum MRR. They observed that MRR was maximum at increasing T_{on} and decreasing T_{off} while increasing spark voltage provides lower material removal rate. They also explored that MRR was minimum at first level of T_{on} and maximum at first level of T_{off}. For obtaining maximum MRR and minimum SR, the optimum value of parameters was found as spark voltage 30 V, T_{on} 10 μs, T_{off} 2 μs, wire feed 4 m/min and 3% reinforcement [23].

An investigation was carried out on Al/ZrO_2 particulate-reinforced MMC material using wire EDM process for the optimization of machining characteristics by Garg et al. The experiments were designed using RSM technique considering full factorial approach. The process parameters such as pulse width, T_{off}, short discharge, voltage, wire feed rate, and wire tension were used for accommodating the output such as velocity of cutting and machined surface quality. The comparison and validation of RSM and GRA techniques were carried out along with the output parameters characteristics using various wire electrodes and found that diffused wire electrode provided better performance results as compared to brass wire electrode. The optimal machining conditions for multi-objective optimization were 1.14 s pulse width, 4.61 μs time between pulses, 0.7 μs short pulse time, 50 V voltage, 11.1 m/min wire feed rate, and 0.43 daN wire tension. Their optimal value of performance parameters was 8.234 mm/min cutting velocity and 1.803 μm surface roughness [24].

3.5 Conclusion

The machining of advanced materials using EDM provides more precise performance that is appraised by SR, MRR, and tool wear. This performance is mainly affected by the working parameters like T_{on}, T_{off}, additive powder, and electrolyte. There are various techniques such as Taguchi, RSM, and GRA used to optimize the performance parameters of EDMed AMC. From the previous work, it is found that GRA is the most appropriate technique among other methods. This article provides a brief information of EDM process along with theoretical and experimental reviews that have been done by researchers. Moreover, this also discussed various optimization techniques which are used to optimize the process parameters of EDM. This article halted, and optimum parameters of EDM accommodated the satisfactory performance parameters such as material removal, surface roughness, and TWR.

References

1. Boothroyd, G., Winston, A.K.: Non-conventional Machining Processes. Fundamentals of Machining and Machine Tools. Marcel Dekker Inc, New York (1989)
2. Singh, C.D.: Electrical Discharge Machining. Optimization of Chromium Powder Mixed EDM Parameters During Machining of H13 Tool Steel. Anchor Academic Publishing (2017)
3. Ho, K.H., Newman, S.T.: State of the art electrical discharge machining (EDM). Int. Journal Mach. Tool. Manuf. **43**(13), 1287–1300 (2003)
4. El-Hofy, H.A.: A Text Book of Advance Manufacturing Processes. Tata Mc Graw Hill (2005)
5. Mc Geough, J.A.: Advanced Methods of Machining, 1st edn. Chapman and Hall, USA (1988)
6. Dilshad, A.K., Mohammad, H.: Effect of tool polarity on the machining characteristics in electric discharge machining of silver steel and statistical modeling of the process. Int. J. Environ. Sci. Technol. **3** (2011)
7. Lonardo, P.M., Bruzzone, A.A.: Effect of flushing and electrode material on die-sinking EDM. CIRP-Annals Manuf. Technol. **48**(1), 123–126 (1999)
8. Wong, Y.S., Lim, L.C., Lee, L.C.: Effect of flushing on electro discharge machined surfaces. J. Mater. Proc. Technol. **48**, 299–305 (1995)
9. Prabu, M., Ramadoss, G., Narender S.P., Christy T.V., Vedhagiri, E.V.: Electrical discharge machining of Al-TiB_2 with low-frequency vibrating tool. Sci. Eng. Com. Mater. **0**, 1–8 (2013)
10. Karthiresan, M., Sornakumar, T.: EDM studies on aluminum alloy-silicon carbide composites developed by vortex technique and pressure die casting. J. Miner. Mater. Charact. Eng. **9**, 77–88 (2010)
11. Nanimina, A. Mouangue, Abdul-Rani, A.M., Ahmad, F., Zainuddin, A., Jason, Lo, S.H.: Effects of electro discharge machining on aluminum metal matrix composite. J. Appl. Sci. **11**, 1668–1672 (2011)
12. Mohan, B., Rajadurai, A., Satyanarayana, K.G.: Effect of SiC and rotation of electrode on electric discharge machining of Al-SiC composite. J. Mater. Proc. Technol. **124**(3), 297–304 (2002)
13. Islam, S., Mark, R.: Electro-discharge machining of metal matrix composite materials. Adv. Mater. Process. Technol. **2**(2), 235–244 (2016)
14. Porwal, R.K., Yadava, V., Ramkumar, J.: Multi objective optimization of hole drilling electrical discharge micromachining processing using grey relational analysis coupled with principal component analysis. J. Inst. Eng. (India): Series C **94**, 317–325 (2013)
15. Bhuyan, R.K., Shalini, M., Routara B.C.: RSM and fuzzy logic approaches for predicting the surface roughness during EDM of Al-SiCp MMC. In: The 5th International Conference of Materials Processing and Characterization (2016)
16. Radhika, N., Kishore, C.G., Shivaram, P., Vijay, K.T.: Multi-Objective optimization of EDM parameters using grey relational analysis. J. Eng. Sci. Technol. **10**(1), 1–11 (2015)
17. Prasanna, P., Tallapragada, V.S., Sashank, S.P., Manikanta, B., Pruthvi, A.: Optimizing the process parameters of electrical discharge machining on AA7075-SiC alloys. Mater. Today: Proc. **4**(8), 8517–8527 (2017)
18. Anand, S., Mohinder, P.G., Kapil, K.G.: Prediction of optimal conditions for WEDM of Al 6063/ $ZrSiO_4$ (p) MMC. Procedia Mater. Sci., 1024–1033 (2014)
19. Senthil, P., Vinodh, S., Adarsh, K.S.: Parametric optimization of EDM on Al-Cu/TiB_2 metal matrix composites using TOPSIS method. Int. J. Mach. Mach., Mater (2014)

20. Wang, C.C., Yan, B.H.: Blind-hole drilling of Al_2O_3/6061Al composite using rotary electro-discharge machining. J. Mater. Process. Technol. **102**, 90–102 (2000)
21. Ming, W., Ma, J., Zhang, Z., Huang, H., Shen, D., Zhang, G., Huang, Y.: Soft computing models and intelligent optimization system in electro-discharge machining of SiC/Al composites. Int. J. Adv. Manuf. Technol. **87**(1–4), 201–217 (2016)
22. Kansal, H.K., Singh, S., Kumar, P.: An experimental study of the machining parameters in powder mixed electric discharge machining of Al-10%SiCp metal matrix. Int. J. Mach. Mach. Mater. **1**(4), 396–411 (2006)
23. Udaya, P.J., Moorthy, T.V., Peter, J.M.: Experimental investigations on machinability of Aluminum alloy A413/flyash/B_4C Hybrid composites using wire EDM. In: International Conference of Design and Manufacturing (2013)
24. Sanjeev, K.G., Alakesh, M., Ajai, J.: Multi objective optimization of machining characteristics during wire electrical discharge machining of Al/ZrO_2 particulate reinforced metal matrix composite. J. Eng. Res. **1**(3), 145–160 (2013)

Chapter 4
Wear Characteristics of Untreated and Alkali-Treated Rice Husk–Epoxy Bio-composite

Neeraj Bisht and P. C. Gope

Abstract Use of bio-composites is gaining acceptance in the engineering world swiftly. However, materials wear out and erode with passage of time. Not much study, however, has been done on this aspect of bio-composites. In this paper, study of tribological behaviour of rice husk-reinforced composites has been done. The effect of rice husk addition has been observed on the wear characteristics of the composite using a pin-on-disc machine. It is seen that the increase in weight percentage of rice husk results in poorer wear characteristics. The SEM micrographs showed improper adhesion between the fibre and the matrix as the reason behind this deterioration in wear strength. Pre-treatment of rice husk with sodium hydroxide improves the adhesion between the matrix and the fibre which is evidently proven by the improvement in wear rate. The increase in wear strength is further corroborated by SEM photographs. It can be concluded that improving the adhesion between the fibre and the matrix by alkali treatment can enhance the working life of the material which can be beneficial in economic and environmental terms.

Keywords Rice husk · Epoxy · Treatment · Wear rate

4.1 Introduction

The worldwide production of rice is approximately 700 million tonnes [1], about 22% of which is rice husk [2]. The rice husk left after milling has no potential application and is mostly used for burning which is not feasible due to the growing environmental concerns; besides, the ash remaining after burning is difficult to dispose. However, rice husk can be a cheap source of reinforcing material either in its raw form or by-products of its thermal degradation. Synthetic fibres are being replaced by natural fibres, are lighter, can be replaced at a higher level, are economical, are biodegradable

N. Bisht (✉) · P. C. Gope
Department of Mechanical Engineering, G.B. Pant University of Agriculture and Technology, Pantnagar, India
e-mail: neerajbisht30@gmail.com

I. Singh et al. (eds.), *Trends in Materials Engineering*,
Lecture Notes on Multidisciplinary Industrial Engineering,
https://doi.org/10.1007/978-981-13-9016-6_4

and are less corrosive for the environment among others (sometimes free) [3, 4]. In the past decade, rice husk in its various forms has been used by scientists in the development of bio-composites. There have been numerous studies regarding the mechanical behaviour of rice husk-reinforced composites. Rice husk has also been utilized as chopped rice husk [5]; however, researchers have mostly used it in a powdered form. Different polymers have been used as matrix material. RH was incorporated in polypropylene by Yang et al. [6], Premalal et al. [7], Dimzoski et al. [8], Rosa et al. [9] to name a few. Besides, polypropylene rice husk has also been used with other polymeric materials. Polyethylene has been utilized by Ghofrani et al. [10], Atuanya et al. [11], Rahman et al. [12], Nawadon et al. [13]. Rubber as a potential matrix material was investigated by Attharangsan et al. [14] and Ramasamy et al. [15]. Besides these polyurethane [16], resin [17], polylactic acid [18], vinyl ester [19], tires [20], etc., have also been used as matrix materials. The incorporation of rice husk, however, has been mostly detrimental to the mechanical strength of the composites. It has been seen that the adhesion between rice husk and polymers is not good due to the presence of waxes and lignin on the surface of the fibre. Certain surface modifications have, however, proved to be effective in promoting adhesion.

Various techniques of surface modification of rice husk have been used till now. The methods can be categorized as use of compatibilizers [13, 21], mercerization [22], plasma treatment [22], electron beam irradiation [23] and treatment with other chemicals [24–28]. All the techniques mentioned above contribute to the improvement in adhesion between the matrix and the fibre.

However, most of these chemical modifications have concentrated on tensile strengths and impact and flexural strengths, and there are very few studies on the impact of surface treatment on the wear characteristics. The result in loss of material from the surface due to interactions of a solid surface's exposed part with interfacing materials and environment can be called wear. Estimated direct and consequential annual loss to industries in the USA due to wear is approximately 1–2% of GDP [29]. Abrasive wear measurements under multi-pass condition on a single pin-on-disc machine were conducted by Chand et al. [30] on composites prepared by RH and polyvinyl chloride. It was seen that as the sliding distance increases the specific wear rate decreases. It was also observed that with increasing reinforcement the specific wear rate also increases, the reason behind which is the improper adhesion between the filler and the matrix material, PVC in this case. The wear rate of reinforced polymer was even higher than pure polymer for all percentages of reinforcement. However, the modification of rice husk with maleic anhydride was seen to be beneficial in terms of wear strength with wear strength increasing with modification; however, the wear strength was still significantly higher for higher filler ratio.

Majhi et al. [31] studied the tribological behaviour of RH-filled epoxy composites. It was observed that with increasing sliding distance the wear rate decreased. The effect of increasing the reinforcement was that the wear rate first decreases with increasing fibre content compared to pure epoxy for up to 10% RH loading; thereafter, it started increasing with further increase in filler loading. Same was the case with specific wear rate. Modification of rice husk with benzoyl chloride increased the compatibility between the RH and the polymer matrix, and as a result the wear rate and specific wear rate both are seen to be decreasing.

Chand et al. [32] in another study have examined the wear character of jute fibre-based composite. Addition of maleic anhydride-grafted polypropylene was also studied, and it was seen that the addition of the compatibilizer improved the wear characters of the bio-composite. Keeping in view the fact that the wear characteristics of bio-fibre-reinforced polymer composites were poor, the study is conducted to study the effect of chemical modifications (alkali treatment) on the wear characteristics of these classes of composites.

4.2 Material Preparation

4.2.1 Materials

CY-230 and HY-951 purchased from M/s Excellence Resins Limited, India, has been used as matrix material. Rice husk was used as the reinforcing agent. The rice husk obtained from mill was thoroughly washed with tap water to remove any foreign impurities and sun dried to remove moisture; thereafter, it was grounded and separated on the basis of their sizes with the help of sieve and the fine powder obtained with the sieve of ASTM 120 (125 microns) was used for composite fabrication.

4.2.2 Alkali Treatment of Rice Husk

Rice husk flour (RHF) prepared from rice husk was soaked in NaOH solutions of different concentrations at room temperature and was allowed to stand for about 4 h after which RHF was filtered and washed thoroughly to remove NaOH completely with tap water. The NaOH concentrations taken in the study are 2, 4, 6, 8 and 10%. After completely removing NaOH, RHF was dried in an oven at 105 °C for 3–5 h to remove the moisture.

4.2.3 Preparation

Moulds were prepared using 12-mm-thick perspex sheets. Mechanical stirrer was used to mix different weight percentages (wt%) of RH fibre (10, 20, 30 and 40 wt%) and epoxy resin. The solution obtained by mixing of RHF in resin is kept in the furnace at a temperature of 90 $\pm$ 10 °C for two hours. After two hours, the whole solution is taken out and allowed to cool to a temperature of 45 °C. When a temperature of 45 °C is attained, the hardener HY-951 (9 wt%) is mixed immediately [33]. Due to addition of hardener, high viscous solution is obtained which is remixed mechanically by the mechanical stirrer. The viscous solution is then poured into different moulds for sample preparation.

4.2.4 Tribological Testing

The machine used for conducting wear test is Plinth Wear & Friction Machine which basically falls in the category of "pin and disc machines". The machine is based on a lathe belt driven by a 750 W 3-phase motor and having 14 spindle speeds ranging from 25 to 2150 rev/min. The sample pins of diameter 8 and 53-mm length were prepared on the lathe machine and are rotated against a hardened steel disc of diameter 120 mm. The diameter of revolution of the pins is 43 mm.

4.2.5 Wear Rate Measurement

The tests were conducted at a speed of 237 revs/min and load of 10 N. The pins were rotated against the steel disc for different numbers of revolutions (2000, 4000, 6000, 8000, 10000), and corresponding sliding distance was calculated according to the formula

$$\text{Sliding distance } x = 0.251n = 0.00419Nt$$

where

n	total revolutions
N	disc speed, rev/min
T	duration of run
Radius of test track	40 mm

After each revolution, the weight loss was recorded on a digital weight measuring machine of least count and the specific wear rate was calculated according to the formula

$$K_0 = \frac{W}{\rho L D} \tag{4.1}$$

where W = weight loss in kg, ρ is density in kg/m^3, L is the load in Newton and D is sliding distance in metres.

4.2.6 SEM Studies

Morphological studies of worn surfaces have been done. The images are obtained through microscopic investigation with LEO435V6. To obtain the scanning electron micrographs, square samples are cut from the cast material, gold coated to avoid the artefacts associated with sample charging and then placed inside a chamber in which an electron beam falls on the material. The accelerated voltage was 10 kV. Different images are taken at various magnification ranges.

4.3 Results and Discussion

Figure 4.1 shows the variation of specific wear rate with the weight percentage of rice husk. From the figure, it can be observed that addition of rice husk to the epoxy resin results in increase in specific wear rate which is due to the poor interfacial adhesion between the fibre and the matrix material. The reduction in wear strength can also be attributed to the fact that rice husk has a tendency of agglomeration, thereby making the adhesion improper. It is, however, observed that the specific wear rate decreases with increasing sliding distance due to multi-pass conditions. It has been observed that as the rice husk content increases there is a decrease in hardness and specific wear rate indication, a positive correlation between both the properties, viz. hardness and wear strength.

It can be seen that addition of 10 wt% rice husk has a little impact on the wear strength; however, further increase in rice husk content significantly decreases the wear strength. However in a different research [34] by the same authors, it has been statistically proved that 10 and 20 wt% groups are statistically insignificant in terms of flexural and tensile strengths.

Keeping in mind the advantage of replacing maximum epoxy with rice husk, we have optimized the treatment parameter for 20 wt% rice husk reinforcement. For this,

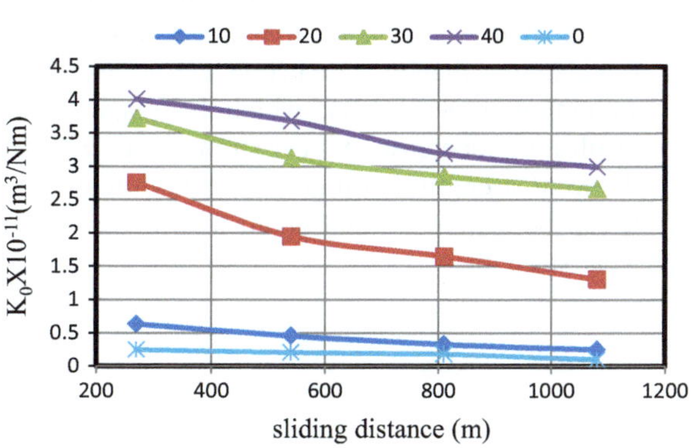

Fig. 4.1 Variation of specific wear rate with treated rice husk (20 wt%) and sliding distance

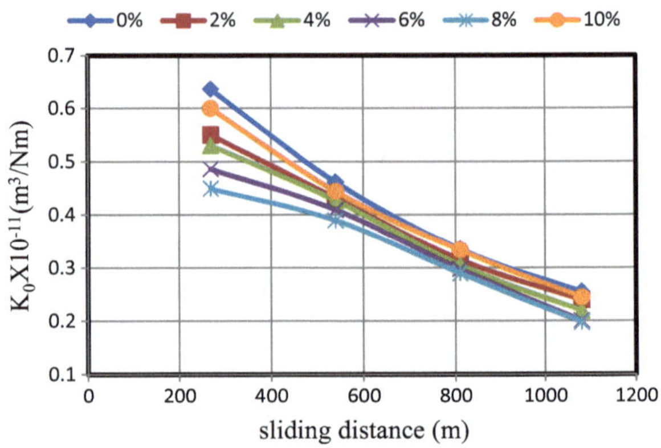

Fig. 4.2 Variation of specific wear rate with treated rice husk (20 wt%) and sliding distance

rice husk was pre-treated with different solutions of 2, 4, 6, 8 and 10% concentrations and wear tests were conducted on them (Fig. 4.2).

From the SEM images of worn surfaces of the samples, it can be observed that as the filler loading increases the surface of the samples becomes rougher. It means that matrix cracking and surface damage are more pronounced. For pre-treated rice husk composite, the surface damage is lesser as compared to untreated samples and it looks like that the fibres are better dispersed in the matrix phase and therefore the wear rate becomes less. Also, it can be seen that for untreated rice husk composites the cracking is in both the longitudinal and the transverse directions, while for pre-treated rice husk the cracking is not as pronounced (Fig. 4.3).

4.4 Conclusions

1. With the increase in rice husk content, the wear strength is seen to be decreasing and at 40% loading it becomes significantly lower than pure epoxy.
2. Pre-treatment with sodium hydroxide improves the adhesion between the fibre and the matrix and results in increase in wear strength.
3. It is also seen that increase in concentration of treatment media enhances the wear strength up to 8% beyond which there is an excessive deterioration of the fibres resulting in decrease in wear strength at 10% concentration solution. However, the wear strength is still greater than untreated rice husk composite.
4. It can be concluded that alkali treatment is an effective technique to increase wear strength.

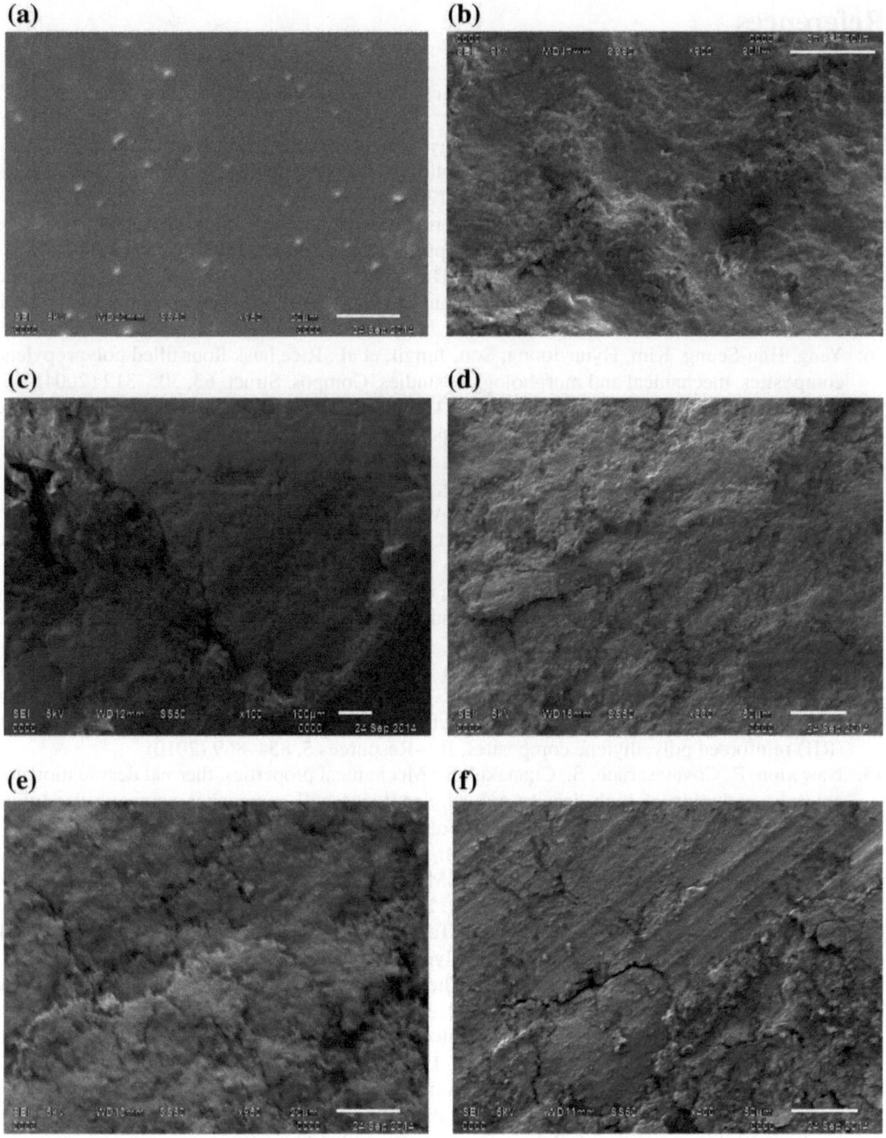

Fig. 4.3 Scanning electron micrograph of worn surfaces of different samples: **a** pure epoxy, **b** 10 wt% rice husk, **c** 20 wt% rice husk, **d** 30 wt% rice husk, **e** 40 wt% rice husk and **f** 8% NaOH-treated rice husk

References

1. Trade and Markets Division. Food and Agriculture Organization of the United Nations, FAO Rice Market Monitoring Bulletin (2010)
2. Beagle, B.C.: Rice-husk conversion to energy: FAO Agricultural Service Bulletin (1978)
3. Ismail, H., Nasir, S.M.: Dynamic vulcanization of rubber wood-filled polypropylene/natural rubber blends. Polym. Testing **20**, 819–824 (2001)
4. Oksman, K., Clemons, C.: Mechanical properties and morphology of impact modified polypropylene–wood flour composites. J. Appl. Polymer Sci. **67**, 1503–1513 (1997)
5. RazaviNouri, M., Jafarzadeh-Dogouri, F., Oromiehie, A., et al.: Mechanical properties and water absorption behaviour of chopped rice husk filled polypropylene composites. Iran. Polym. J. **15**, 757–766 (2006)
6. Yang, Han-Seung, Kim, Hyun-Joong, Son, Jungil, et al.: Rice husk flour filled polypropylene composites, mechanical and morphological studies. Compos. Struct. **63**, 305–312 (2004)
7. Premalal, H.G.B., Ismail, H., Baharin, A.: Comparison of the mechanical properties of the rice husk powder filled polypropylene composites with talc filled polypropylene composites. Polym. Testing **21**, 833–839 (2002)
8. Dimzoski, B., Bogoeva-Gaceva, G., Gentile, G., Avella, M., et al.: Polypropylene-based eco-composites filled with agricultural rice hulls waste. Chem. Biochem. Eng. **23**, 225–230 (2009)
9. Rosa, S.M.L., Santos, E.F., Ferreira, C.A., et al.: Studies on the Properties of Rice Husk filled PP composites-effect of maleated PP. Mater. Res. **12**, 333–338 (2009)
10. Ghofrani, M., Pishan, S., Mohammad, M.R., et al.: A study on rice-husk/recycled high density polyethylene composites—their physical and mechanical properties. Environ. Sci. **9**, 99–112 (2012)
11. Atuanya, U., Olaitan, S.A., Azeez, T.O., et al.: Effect of rice husk filler on the mechanical properties of polyethylene matrix composite. Int. J. Curr. Res. Revisions **05**, 111–118 (2013)
12. Rahman, M.R., Islam, M.N., Huque, M.M., et al.: Effect of chemical treatment on the rice hush (RH) reinforced polyethylene composites, Bio-Resources **5**, 854–869 (2010)
13. Nawadon, P., Covavisaruch, S., Chanakul, S.: Mechanical properties, thermal degradation and natural weathering of high density polyethylene/ rice hull composites compatibilized with maleic anhydride grafted polyethylene. J. Polym. Res. **19**, 1–9 (2012)
14. Attharangsan, S., Ismail, H., Abu Bakar, M., et al.: The effect of Rice husk powder on Standard Malaysian Natural Rubber Grade L (SMR L) and Epoxidized Natural Rubber (ENR 50) composites. Polym. Plast. Technol. Eng. **51**, 231–237 (2012)
15. Ramasamy, S., Ismail, H., Munusamy, Y.: Tensile and morphological properties of rice husk powder filled natural rubber latex foam. Polym. Plast. Technol. Eng. **51**, 1524–1529 (2012)
16. Rozman, H.D., Ang, L.G., Tay, G.S., et al.: The mechanical properties of rice husk-polyurethane composites. Polym. Plast. Technol. Eng. **42**, 327–343 (2003)
17. Ofem, M.I., Umar, M., Ovat, F.A.: Mechanical properties of rice husk filled cashew nut shell liquid Resin composites. J. Mater. Sci. Res. **1**, 89–97 (2012)
18. Yussuf, A.A., Massoumi, I., Hassan, A.: Comparison of polylactic acid/Kenaf and polylactic acid/rise husk composites: the influence of the natural fibres on the mechanical, thermal and biodegradability properties. J. Polym. Environ. **18**, 422–429 (2010)
19. Shivappa, D., Ananda, G.K., Shivakumar, N.: Mechanical characterization of rice husk flour reinforced vinylester polymer composite. Int. J. Innov. Res. Sci. Eng. Technol. **2** (2013)
20. Garcia, D., Lopez, J., Balart, R., et al.: Composites based on sintering rice husk-waste tire rubber mixtures. Mater. Des. **28**, 2234–2238 (2007)
21. Yang, H.-S., Kim, H.-J., Park, H.-J., et al.: Effect of compatibilizing agents on rice-husk flour reinforced polypropylene composites. Compos. Struct. **77**, 45–55 (2007)
22. Nguyen, M.H., Kim, B.S., Ha, J.R., et al.: Effect of plasma and NaOH treatment for rice husk/PP composites. Adv. Compos. Mater. **20**, 435–442 (2011)
23. Ahmad, I., Lane, C.E., Mohd, D.H., et al.: Electron-beam-irradiated rice husk powder as reinforcing filler in natural rubber/high-density polyethylene (NR/HDPE) composites. Compos. B **43**, 3069–3075 (2012)

24. Rozman, H.D., Lee, M.H., Kumar, R.N., et al.: The effect of chemical modification on rice husk with Glycidyl Methacrylate on the mechanical and physical properties of rice husk-polystyrene composites. J. Wood Chem. Technol. **20**, 93–109 (2000)
25. Santiagoo, R., Ismail, H. Hussin, K.: Effects of acetic anhydride on the properties of polypropylene (PP)/recycled acrylonitrile butadiene (NBRr)/rice husk powder (RHP) composites. Polym. Plast. Technol. Eng. **51**, 1505–1512 (2012)
26. Huang, Gu: Tensile behaviours of the coir fibre and related composites after NaOH treatment. Mater. Des. **30**, 3931–3934 (2009)
27. Rout, J., Misra, M., Tripathy, S.S., et al.: The influence of fibre treatment on the performance of coir-polyester composites. Compos. Sci. Technol. **63**, 1303–1310 (2001)
28. Rahmana, M.M., Khan, M.A.: Surface treatment of coir (*Cocos nucifera*) fibres and its influence on the fibres physico-mechanical properties. Compos. Sci. Technol. **67**, 2369–2376 (2007)
29. Zum Gahr, K.-H.: Microstructure and Wear of Materials, 10th edn. Tribology Series. Elsevier Publishers
30. Chand, N., Sharma, P., Fahim, M.: Tribology of maleic anhydride modified rice-husk filled polyvinylchloride. Wear **269**, 847–853 (2010)
31. Majhi, S., Samantarai, S.P., Acharya, S.K.: Tribological behaviour of modified rice husk filled epoxy composite. Int. J. Sci. Eng. Res. **3**(6) (2012)
32. Chand, N., Dwivedi, U.K.: Effect of coupling agent on abrasive wear behaviour of chopped jute fibre-reinforced polypropylene composites. Wear **261**, 1057–1063 (2006)
33. Singh, V.K., Gope, P.C.: Silica-styrene-butadiene rubber filled hybrid composites: experimental characterization and modeling. J. Reinf. Plast. Compos. **29**, 2450–2468 (2010)
34. Neeraj, B., Gope, P.C.: Mechanical properties of rice husk flour reinforced epoxy bio-composite. Int. J. Eng. Res. Appl. **5**(6), 123–128 (2015)

Chapter 5
Tribological Performances of Woven Carbon Fabric/Epoxy Composites Under Dry and Oil Lubrication Condition: An Experimental Investigation

Santosh Kumar and K. K. Singh

Abstract Tribological performance of carbon fiber reinforced epoxy polymer composites (CFRP) in dry and oil lubricating condition. The tests were conducted on a pin-on-disk machine under different applied loads and at a constant sliding velocity. CFRP composite laminates were manufactured by wet lay-up technique followed by vacuum bagging. All the samples were rubbed against a counter-surface of steel (En-31) having hardness of 60 HRC and average surface roughness (R_a) of 0.3 μm. Tribological performances in terms of wear and frictional characteristics were evaluated on the basis of total weight lost during experimentation and specific wear rate of the material and coefficient of friction, respectively. The experimental results show that maximum specific wear rate of CFRP composites with a value of 1.56×10^{-5} mm^3/Nm and 2.77×10^{-5} mm^3/Nm under oil lubricated and dry sliding condition, respectively. The eroded surfaces of CFRP coupons were characterized by Field Emission Scanning Electron Microscope (FESEM) to find out wear mechanism under dry and oil condition.

Keywords Carbon fabric · Wear · Friction

5.1 Introduction

Fiber-reinforced polymer (FRP) composite offers various properties not limited to high-specific strength, specific modulus, internal vibration damping, and superior wear resistance combined with low density. Because of their unique properties, it has been extensively used as advanced composite material [1, 2]. Carbon fiber-reinforced polymer (CFRP) composite exhibits outstanding mechanical as well as tribological properties and is largely used in the field of aerospace, transportation, automotive, biomedical, sporting goods, and mining industries. There are many wear situation encountered by FRP composite materials such as pumps handling industrial fluid,

S. Kumar (✉) · K. K. Singh
Department of Mechanical Engineering, IIT (ISM), Dhanbad, Jharkhand, India
e-mail: kr.santosh.mech@gmail.com

I. Singh et al. (eds.), *Trends in Materials Engineering*,
Lecture Notes on Multidisciplinary Industrial Engineering,
https://doi.org/10.1007/978-981-13-9016-6_5

gears, sewage and abrasive contamination fluids, seals and bushes in agricultural and mining equipment [3–5]. During the past three decades, investigations based on the friction and wear characterization of FRPs are extensively done under dry sliding condition and found that excellent improvement in tribological properties. Nevertheless, CFRP composites exhibit high specific strength, excellent thermal stability and conductivity, and potentially lubricating ability due to their laminated appearance [6]. Vishwanath et al. [7] carried out the friction and wear test for three different woven roving glass fabrics reinforced polyvinyl-butyral-modified phenolic composite under dry sliding condition. Adhesive wear test was performed against a cast iron counter-face. The results reveal that wear rate strongly influenced by glass fabric geometry. Bijwe et al. [8] also studied the influence of three different weave carbon fabric reinforced polyetherimide composites slid against mild steel disk in dry wear mode. They were found that twill composites had better performance. Suresha et al. [9] performed experiments to assess the tribological behavior of carbon/epoxy and glass/epoxy laminates, respectively, against a hard steel disk under dry sliding condition. The results revealed that CFRP composites possess better wear resistance compared to its glass fiber counterparts, irrespective of the speed and applied load. Tribological aspects such as sliding distance, applied load, and velocity are found to be important parameters affecting friction and wear behavior of FRP composite materials. Therefore, tribological behaviors of FRP composites are studied in order to enhance such properties. In general, many researchers were investigated friction and wear behavior for polymer composite under dry sliding condition. However, very few studies have been reported for tribological performance of FRP composite under oil lubricated condition [10–13]. Zhang et al. [14] inspected the tribological characteristics of neat epoxy and epoxy-based composites based on their friction and wear behavior. Such properties were evaluated against stainless steel under diesel lubricated condition. Test result showed that incorporation of short glass or carbon fiber in epoxy improved the tribological properties. Chen et al. [15] investigated wear and frictional properties of carbon/phenolic-resin laminates against stainless steel. The tests were carried out under dry as well as pure and seawater sliding condition. Results reveal that composite had the highest specific wear rate under pure water followed by under dry and seawater. Sandeep Agrawal et al. [16] demonstrated that in GFRP laminates minimum wear rate is found under oil lubricated environmental condition and maximum wear rate found under inert gas sliding environmental condition.

In this study, the comparative investigation of tribological characteristics of woven CFRP composites subjected to dry and oil lubricated sliding conditions.

5.2 Experiment

5.2.1 Material Specification

Epoxy resin (L-12) is well-known thermosetting polymer used as matrix material and K-6 hardener was added for room temperature curing. Epoxy possesses good mechanical as well as tribological properties due to cross-linked polymer and also has low-cost polymer as compared to others. Epoxy (L-12) resin and curing hardener (K-6) were supplied by Atul Industries Pvt. Ltd., India. Bidirectional woven carbon fabrics of 600 GSM were used as reinforcement in epoxy resin.

5.2.2 Material Fabrication

Wet lay-up technique further assisted by vacuum bagging was used to fabricate CFRP laminates. Eight layers of carbon fabric were taken to obtain 4 mm thickness for laminates fabrication. Each of the samples was symmetric laminates with stacked sequence of [(0°/90°)/(−45°/+ 45°)/(−45°/+ 45°)/(0°/90°)/(0°/90°)/(−45°/+ 45°)/(−45°/+ 45°)/(0°/90°)]. Firstly, epoxy and hardener were mixed in the ratio of 10:1. The solution was then applied on the woven carbon fabric using a brush to accomplish the desired stacking sequence. The excess resin was removed using a roller after application of solution to each layer. For further removal of excess resin, vacuum bagging was used at the pressure of 650 mm of Hg for 35 min. After this, a load of 40 kgf was applied for 24 h. Samples of dimensions 8 mm × 8 mm were cut from the laminate composites were glued to the aluminum pins of diameter 6 mm (Fig. 5.1).

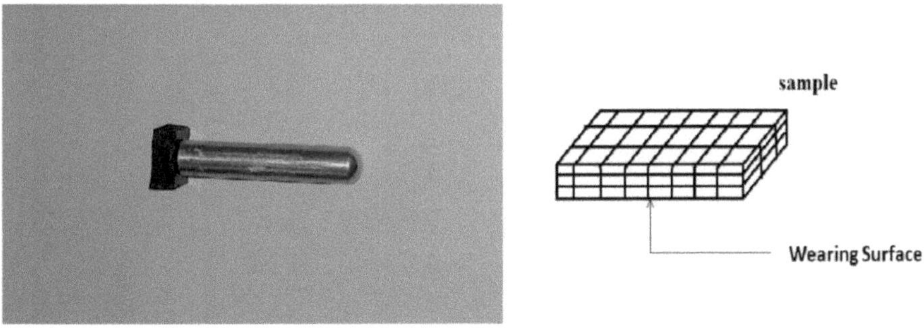

Fig. 5.1 Test sample glued with Al pin

5.2.3 Tribological Testing

The tribological tests were performed on pin on disk (TR-20) by DUCOM. The operating parameters which were taken as follow: sliding velocity 2.61 m/s; variable applied load (25, 50, 100 N); sliding time, 5 min and sliding distance, 0.786 km. In this study, the experiment was performed in dry and oil environment condition at room temperature. SAE20 oil was continuously supplied through elastomer tube onto the wear track under gravity. En-31 steel disk with surface hardness of 60 HRC and R_a (average surface roughness) of 0.3 μm was used counterpart. Each specimen was weighted before and after the experiment. The sliding direction of the specimen fibers was kept parallel and anti-parallel.

5.3 Results and Discussion

5.3.1 Weight Loss and the Specific Wear Rate

The weight loss and specific wear rate (K_0) of CFRP composite samples were calculated with different applied load (25, 50 and 100 N) and others parameters kept constant as discussed above in dry and oil lubricated condition. Weight loss under different normal load for different samples under dry and oil lubricated condition is displayed in Fig. 5.2a. It is observed that weight loss is increasing with increase of applied loads for both lubricating conditions. K_0 is one of the factors to determine the tribological characteristics more precisely as compared to material mass loss. The specific wear rate of a material is calculated as;

$$\text{Specific Wear Rate}(K_0) = \frac{\Delta m}{\rho \times L \times d}\left(\frac{\text{mm}^3}{\text{N} - \text{m}}\right) \tag{1}$$

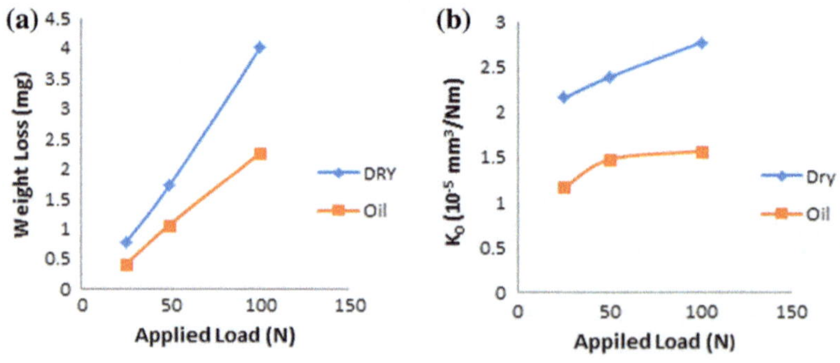

Fig. 5.2 Graphs depicting **a** Weight loss versus applied load; and **b** Specific wear rate (K_0) under different environmental condition at a constant velocity of 2.61 m/s

where

Δm mass loss (kg)
ρ density of the composite sample (kg/mm^3)
L normal applied load (N)
d sliding distance (m).

From Fig. 5.2b, it may be observed that with the increment in normal applied load, the value of specific wear rate increases. It is obvious that with the increment in load, weight loss increases. From the above formula, it may be noted that both numerator (weight loss) and denominator (load) increases with the increment in normal applied load. So as per the observations, the increased value of specific wear rate determines that rate of increment of weight loss is higher than that of change in value of normal applied load. Nevertheless, the specific wear rate is lower in oil lubrication as compared to dry is due to boundary lubrication formation between the pair surfaces.

5.3.2 *The Friction Coefficient*

Coefficient of friction is used to analyze the friction force between mating surface and frictional behavior of composite materials. Friction results from the asperities at micro-level. Figure 5.3 shows that with increasing load, frictional coefficient increases. Common phenomenon that governs the friction behavior are either increment in the amount of carbon fibers on disk surface in the form of wear debris that acts as solid lubricant or the formation of a tribo-film that reduce the coefficient of friction between the contacting surface. Moreover, after prolonged mating between tribo-surfaces formation of new surface due to removal of previous tribo-surface is inevitable. The formation of new surfaces with different tribo properties leads to uneven profile of new surface that result in higher coefficient of friction. From the experimental result, it was found that the normal applied load has a direct effect on

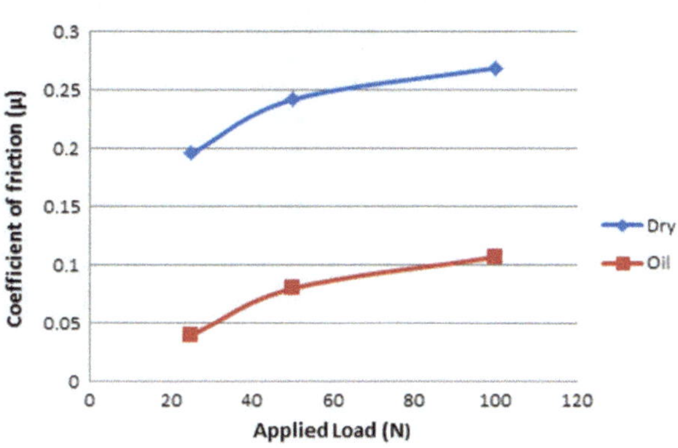

Fig. 5.3 Variation in frictional coefficient under various applied load at v = 2.61 m/s

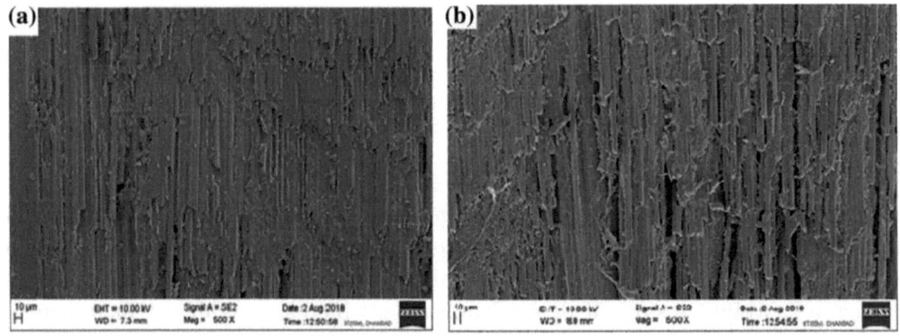

Fig. 5.4 FESEM image of worn out surfaces composite sample under dry condition, **a** 25 N, 2.16 m/s and **b** 100 N, 2.16 m/s

the frictional coefficient which being an increase in the load increase the friction or vice versa.

5.3.3 *FESEM Analysis*

FESEM examination of worn surfaces CFRP composite samples under normal applied load 25 and 100 N and at 2.61 m/s constant sliding velocity are shown in Figs. 5.4a, b and 5.5a, b. There are four wear mechanisms mainly involved under sliding wear condition for fiber-reinforced polymer composites [17]. It has been also observed from FESEM images that four wear mechanisms involved under dry and lubricated condition. Figure 5.4a, b show worn out surface features of carbon fiber-reinforced epoxy composite samples under normal sliding condition. It could be visibly seen that there is maximum materials removal occurred through fiber wear,

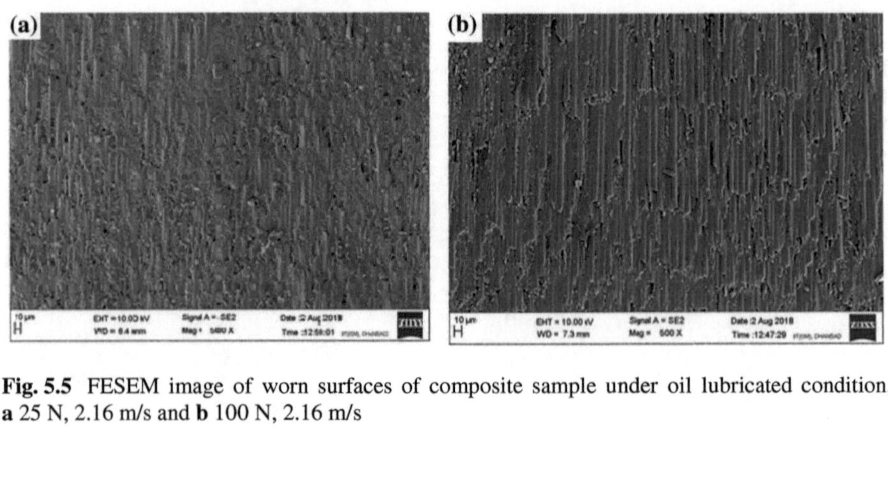

Fig. 5.5 FESEM image of worn surfaces of composite sample under oil lubricated condition, **a** 25 N, 2.16 m/s and **b** 100 N, 2.16 m/s

matrix wear, severe fiber fracture, and interfacial debonding due to higher heat generated at the interface. Figure 5.5a, b show worn surface of carbon fiber-reinforced epoxy samples under oil lubricated condition, in which lower fiber wear and matrix wear observed and very few fibers debonding happened.

5.4 Conclusion

Tribological characterization of CFRP laminates conducted in this work could be concluded as;

1. Wear as well as frictional characteristics of CFRP laminates is highly influenced by sliding medium and applied loads as compared to others parameters.
2. Under oil lubrication conditions, coefficient of friction was found to be minimum because of the induction of thin oil layer leading to lubricating interface of tribo-pair.
3. Specific wear rate and weight loss of composite increases with increasing load normal to the specimen irrespective of the sliding conditions.
4. FESEM observation of worn surfaces shows wear mechanisms involved during wear process are fiber breakage, matrix wear, and fiber debonding.

References

1. Mallick, P.K.: Fiber Reinforced Composites: Materials, Manufacturing and Design, 3rd edn. Taylor & Francis, Boca Raton, Florida (2008)
2. Bhushan, B.: Introduction to Tribology, 2nd edn. Wiley, Ohio (2013)
3. Friedrich, K., Lu, Z., Hager, A.M.: Recent advances in polymer composites' tribology. Wear **2**(190), 139–144 (1995)
4. Harsha, A.P., Tewari, U.S., Venkatraman, B.: Three-body abrasive wear behaviour of polyaryletherketone composites. Wear **7–8**(254), 680–692 (2003)
5. Suresha, B., Kumar, K.N.S.: Investigations on mechanical and two-body abrasive wear behaviour of glass/carbon fabric reinforced vinyl ester composites. Mater. Des. **6**(30), 2056–2060 (2009)
6. Lee, H.G., Hwang, H.Y., Lee, D.G.: Effect of wear debris on the tribological characteristics of carbon fiber epoxy composites. Wear **3–4**(261), 453–459 (2006)
7. Vishwanath, B., Verrnab, A.P., Kameswara Rao, C.V.S.: Effect of reinforcement on friction and wear of fabric reinforced polymer composites. Wear **167**, 93–99 (1991)
8. Bijwe, J., Rattan, R.: Influence of weave of carbon fabric in polyetherimide composites in various wear situations. Wear **7–12**(263), 984–991 (2007)
9. Suresha, B., Chandramohan, G., Samapthkumaran, P., Seetharamu, S., Vynatheya, S.: Friction and wear characteristics of carbon-epoxy and glass-epoxy woven roving fiber composites. J. Reinf. Plast. Compos. **2**(25), 771–782 (2006)
10. Zhang, Z., Liu, W., Xue, Q., Shen, W.: Friction and wear characteristics of PTFE composites filled with metal oxides under lubrication by oil. J. Appl. Polym. Sci. **1**(66), 85–93 (1997)
11. Zhao, F., Li, G., Osterle, W., Lnes, H., Zhang, G., Wang, T., Wang, Q.: Tribological investigations of glass fiber reinforced epoxy composites under oil lubrication conditions. Tribol. Int. **103**, 208–217 (2016)

12. Agrawal, S., Singh, K.K., Sarkar, P.K.: Comparative investigation on the wear and friction behaviors of carbon fiber reinforced polymer composites under dry sliding, oil lubrication and inert gas environment. Mater. Today: Proc. **1**(5), 1250–1256 (2018)
13. Fei, J., Li, H.J., Huang, J.F., Fu, Y.W.: Study on the friction and wear performance of carbon fabric/phenolic composites under oil lubricated conditions. Tribol. Int. **56**, 30–37 (2012)
14. Zhang, G., Burkhart, T., Wetzel, B.: Tribological behavior of epoxy composites under diesel-lubricated conditions. Wear **307**, 174–181 (2013)
15. Chen, B., Yang, J., Wang, J., Liu, N., Yan, F.: Comparative investigation on the friction and wear behaviors of carbon fabric-reinforced phenolic composites under seawater lubrication. Tribol. Trans. **1**(58), 140–147 (2015)
16. Agrawal, S., Singh, K.K., Sarkar, P.K.: A comparative study of wear and friction characteristics of glass fibre reinforced epoxy resin, sliding under dry, oil-lubricated and inert gas environments. Tribol. Int. **96**, 217–224 (2016)
17. Friedrich, K.: Microstructural efficiency and fracture toughness of short fiber/thermoplastic matrix composites. Compos. Sci. Technol. **1**(22), 43–74 (1985)

Chapter 6
Effect of Stacking Sequence and Fiber Volume Fraction on the Static Mechanical Properties of Woven GFRP Composite

Md. Touhid Alam Ansari, K. K. Singh and Md. Sikandar Azam

Abstract This paper evaluates the effect of fiber volume fraction and ply stacking sequence on the tensile and flexure strength of GFRP laminates. Six- and 8-plied laminates of equal thickness with two different ply stacking sequences (symmetric and asymmetric laminates) and different fiber volume fractions ($v_f = 45$ and 56%) were fabricated using press molding method. Flexural strength was evaluated under a 3-point bend test according to ASTM D790 and tensile test as per the ASTM D3039 testing standard. Results showed that the static mechanical properties of the laminates were a function of fiber volume fraction and fiber orientation. Samples with high v_f presented low tensile strength and high flexure strength compared to the specimen with lower v_f. While among symmetric and asymmetric stacking sequence, symmetric one independent of v_f had better tensile and flexure properties.

Keywords GFRP · Stacking sequence · Flexural strength

6.1 Introduction

Higher specific strength along with better tailoring properties of the glass fiber-reinforced polymers (GFRP) laminates makes it one of the most favored materials in the area of the automobile, aviation, wind turbine, and sports. However, in order to have a longer life, high fatigue life is highly anticipated [1]. GFRP compared to its carbon fiber counterpart; i.e. CFRP shows poor fatigue life and high strength degradation [2]. Glass fiber-reinforced polymer (GFRP) composites are most commonly used in the manufacture of composite materials as woven fibers have the extra benefit of through-thickness reinforcement and has similar properties in both warp and weft directions [3].

Md. T. A. Ansari (✉) · K. K. Singh · Md. S. Azam
Department Mechanical Engineering, Indian Institute of Technology (Indian School of Mines), Dhanbad 826004, Jharkhand, India
e-mail: touhidalam90@gmail.com

I. Singh et al. (eds.), *Trends in Materials Engineering*,
Lecture Notes on Multidisciplinary Industrial Engineering,
https://doi.org/10.1007/978-981-13-9016-6_6

The laminate properties are influenced by two major features, i.e., material properties and geometry [4]. When the plies stacking sequence of laminates are mirror image of each other about mid plane, the stacking leads to symmetric laminates; if not then asymmetric one. Both laminate sequences represent different material properties regarding stiffness matrix and variation in mechanical response under loading. Thus, the study of variation in mechanical properties is an important part of the laminate design and its application should be discussed. Nevertheless, Singh et al. [5] discussed the stiffness characteristics of symmetric and asymmetric FRP laminates and found that the stiffness matrices of both the laminates are almost identical. However, in spite of possessing better tenacity, these materials are susceptible to damage because of their low interlaminar strength [6].

To determine flexural strength, a 3-point bend test is conducted in which a loading roller bends a sample at a fixed span length and loading rate, until rupture. Once the load is applied, the test sample bends in such a way that the upper side goes under compression while the bottom of the test sample is subjected to tension [7]. Sample also experiences shear stress along the mid-plane. Particularly because of the anisotropic behavior of the GFRP samples, bending failure may be produced by individual tensile, compressive and shear stresses or their combination. Nunes et al. studied the relevance of a 3-point bend test to calculate the stiffness of anisotropic composite laminates in bending. They found that the flexural behavior of the composites is influenced by many factors, including laminate stacking, fiber orientation, cure behavior, fiber volume fraction and surface waviness [8]. Many researchers have studied the phenomena of damage initiation and growth in composite structures subjected to bending loads using various FE models [9, 10]. Dong et al. experimentally investigated the flexural properties of bidirectional hybrid epoxy composites reinforced with carbon and glass fiber [11]. However, limited resource in the stipulated field prompted the authors to conduct tensile and flexure failure behavior of these modern marvels.

As mentioned, laminates with two different volume fractions and stacking sequence were considered for these set of experiments. The stress–strain curves were obtained which were used to evaluate the effect of stacking sequence and fiber volume fraction in the composite laminates.

6.2 Materials and Fabrication Method

In the present study, bidirectional plain woven E-glass fabrics having surface weight 600 GSM and epoxy compatible sizing supplied by S & S polymers, Bangalore, and Bisphenol-A based thermosetting epoxy (Araldite LY 556 CS) and hardener (Aradur Hy 951) supplied by the same company were used as matrix material in a ratio of 10:1. As per manufacturer instruction, flow time is 20 min for the epoxy and hardener mixture in the above-mentioned ratio.

Laminate fabrication is done by using the plain woven glass fiber of three types of orientations: (0°/90°), (±45°) and (±60°). Plies having a dimension of $28 \times 28\ cm^2$

Table 6.1 Stacking sequences for laminates

Laminate-Types		Symmetric	Asymmetric
1	8 plied	(A/B/B/A// A/B/B/A)	(A/B/B/A// B/A/B/A)
2	6 plied	(A/C/A//A/C/A)	(A/C/A//C/A/A)

Where A = (0°/90°), B = (±45°), C = (±60°)

were cut from the fiber fabric according to the fiber orientation. Laminates with proposed stacking sequence (shown in Table 6.1) were fabricated. The first layer was placed on a glass surface and epoxy was applied to the plies with a soft brush over it followed by the second layer, and similarly, the plies were placed as per the introduced stacking sequence. Subsequently, using a mild steel roller, plies were rolled to squeeze out the extra epoxy and to eliminate air bubble trapped in between them. Eight- and 6-layered GFRP laminates were prepared by hand lay-up method followed by press molding. Lastly, these laminates were placed in a hydraulic press under a pressure of 0.225 MPa for 24 h for curing. Hydraulic press machine and fabricated laminate are shown in Fig. 6.1. A square bracket of thickness 3 mm is used to maintain the thickness of laminates. Fiber volume fraction (V_f) of 6 and 8 plies were calculated as 45 and 56% approximately according to ASTM D3171-15.

Specimens for tensile strength and flexural strength were prepared as per ASTM D3039 and ASTM D790, respectively. The specimen configuration is shown in Fig. 6.1c, d. For the tensile test, specimens having a gauge length of 100 mm and width 20 mm with 50 mm grip on both ends are used. Taper aluminum end tabs were introduced to remove grip-related failure in the material. Specimen dimension for 3-point bend test is 48 mm × 12 mm × 3 mm. Four samples of each stacking sequence laminate are tested using Hounsfield H50KS computer-controlled universal testing machine (UTM) with 50 KN load cell at a crosshead speed of 1 mm/min under atmospheric condition. The 3-point bend tests were performed to get flexural stress which is calculated by the following equation:

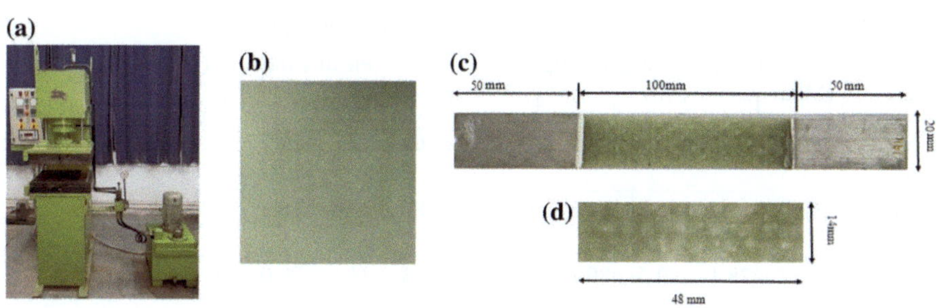

Fig. 6.1 **a** Hydraulic press machine, **b** prepared laminate, **c** tensile test coupon and **d** 3-point bend test coupon

$$\sigma_f = \frac{3PL}{2bd^2}$$

where

- σ_f flexural stress of sample (in MPa)
- P load (in N),
- L length of span (in mm),
- b width of sample (in mm) and,
- d depth of sample (in mm).

6.3 Result and Discussion

6.3.1 Tensile Behavior of GFRP

Ultimate tensile stress and strain were estimated according to ASTM D3039/3039M testing standard. Figure 6.2a shows a stress–strain diagram for the tested samples. The results of the tensile test on the prepared specimen with different stacking sequence are given in Table 6.2. Which were found to be fiber volume fraction dependent. For the similar thickness, 6-plied laminate had a smaller fiber fraction compared to the 8-plied laminate. Thus, the higher tensile strength of 380 MPa was obvious in 6-plied laminate compared to that of 8-plied laminates that stood at 342 MPa. Lack of matrix in 8-plied laminate could be attributed to such results and similar results were presented by Mini et al. [12].

Ply stacking sequence of laminates dictates the tensile strength of the GFRP composite and is well studied by many authors [13, 14]. Samples witnessed mid-axis failure in the gauge length. Upon loading, laminates first witness matrix micro-cracking in 90° plies. Further increase in load propagates these micro-cracks into fiber direction. Similarly, angled plies witness these cracks in the matrix that propagates in

Table 6.2 Result of tensile and flexure test of different stacking sequences

Laminate types	σ_{ut} (MPa)	Flexural strength(MPa)					Flexure modulus(GPa)				
		1	2	3	4	Avg.	1	2	3	4	Avg.
6-plied sym	380.5	464.1	420.3	394.0	380.1	414.63	20.3	19.7	19.4	19.2	19.6
6-plied asym	365.4	415.7	392.7	380.5	362.3	387.80	20.8	18.1	20.3	20.3	19.9
8-plied sym	342.2	478.1	479.6	480.1	483.9	480.43	23.1	26.0	26.2	26.6	25.5
8-plied asym	328.4	425.3	440.5	462.5	466.2	448.63	20.2	17.8	19.4	19.0	19.1

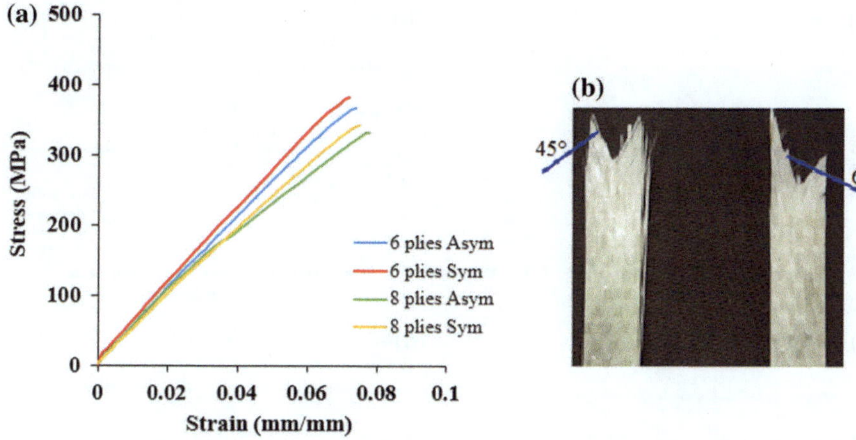

Fig. 6.2 **a** Stress–strain graph of different stacking sequences; **b** failure mode of tensile specimen

fiber direction as well. However, in the absence of matrix, 8-plied laminates obtained matrix crack saturation at a rapid rate compared to 6-plied laminate. Further, these cracks propagated in fiber direction into the fiber–matrix interface and reduced the strength. Ply-intersecting angle dominates the tensile behavior of the materials as well. A small ply-intersecting angle (90°$-\theta$°) between the adjacent plies diminishes inter-ply cracking and hinders crack growth in oblique angle plies. 6-plied laminate has small intersecting laminate compared to 8-plied laminate. This large intersecting angle promoted inter-ply crack growth and coalescence and samples attained lower tensile strength. Figure 6.2b shows the failure pattern of the laminates with different ply angles. .It was observed that final fracture of the samples was dominated by the angle plies, i.e. fibers with (±45) and (±60°) direction failed in mixed mode direction due to the presence of warp and weft direction fibers, resulting in a 'V' shape failure (as shown in Fig. 6.2(b)). In symmetric laminates, an axial load is equally countered and presents better tensile properties, whereas lack of symmetricity in asymmetric laminate fails to counterbalance the forces leading to early failure.

6.3.2 *Flexural Behavior of GFRP*

Samples loaded under 3-point bending witness shear failure because of the stresses reaching the allowable interlaminar shear strength of the material. Under flexure loading samples withstand a combined tensile, compressive and shear failure at the point of loading. Crack propagation in samples under flexure loading depends upon the fiber volume fraction and adjacent ply orientation, since it is a matrix-dominated property. Woven fiber laminates have a tendency to form matrix-rich regions and non-uniform distribution of the same could severely hamper the flexure properties.

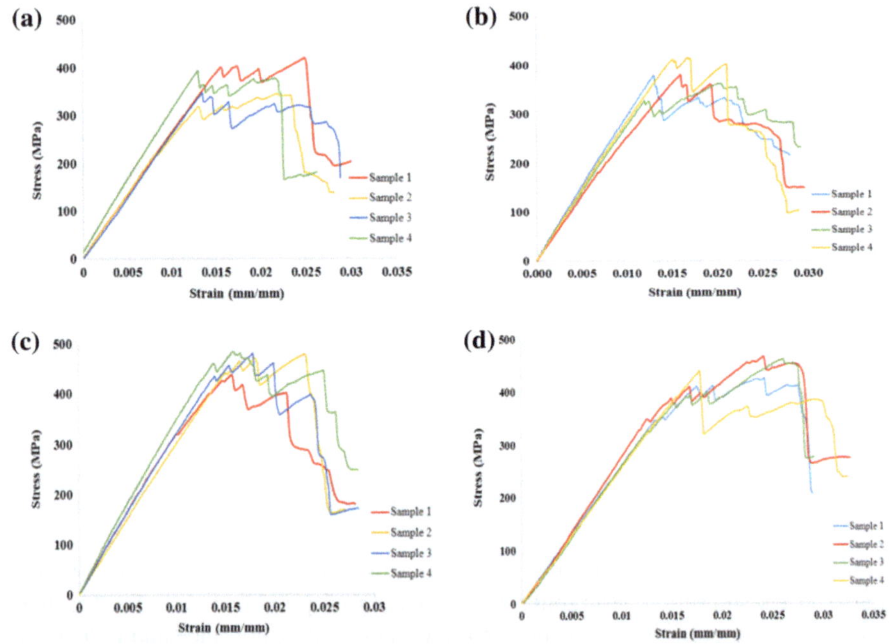

Fig. 6.3 Flexural stress–strain curve for **a** 6-plied symmetric, **b** 6-plied asymmetric, **c** 8-plied symmetric and **d** 8-plied asymmetric laminate

In this work, four different samples with varying fiber volume fraction and fiber orientation were tested, and it was found that 8-plied laminates independent of the stacking sequence presented higher flexure strength, whereas in spite of having high matrix content, 6-plied laminates presented low flexure strength. Flexural stress-strain curves for all four different type of samples are shown in Fig 6.3. The stipulated reason could be the formation of a high number of matrix-rich regions compared to the 8-plied laminates. Matrix-rich regions offer less resistance to the crack propagation compared to the fiber-dominated ones, and 8-plied laminates have higher fiber-dominated regions. Similarly, in laminate with higher fiber content, the fibers are not sufficiently wetted by the epoxy matrix, and the presence of fiber bundles in the form of agglomerates are significantly capable of blocking cracks and consequently stress transfer [15]. Hence, laminates having lower volume fraction fail early when compared to high fiber volume fraction.

In symmetric laminates, a large number of matrix micro-cracking in ±45° and ±60° could be observed when 0°/90° ply cracks initiated and propagated in the fiber direction. The intersecting angle of the cracked plies has significant effects on the process of crack development in the adjacent plies. When the intersecting angle, i.e., (90°−θ°) is small, micro-cracks are observed before propagating the crack in the fiber direction. Ply θ° (i.e., ±45° or ±60°) having a smaller intersecting angle and the thinner thickness leads to the more proneness to micro-crack formation in the adjacent plies [9]. In asymmetric laminates, the $\theta°$ (i.e., $\pm 45°$ or $\pm 60°$) ply becomes thinner as compared to the same ply in symmetric laminates due to the arrangement of plies. Hence, the formation of micro-cracks is more in this case subsequently resulting in the decrease of the flexure strength of asymmetric laminates.

6.4 Conclusion

Through the present work, an attempt was made to experimentally relate the effect of stacking sequence and fiber volume fraction on the tensile and flexure strength of GFRP laminates. Based on the above-stated experimentations, the following conclusions can be stipulated:

1. Laminates with higher fiber content presented lower tensile strength, mostly due to the absence of bulk material.
2. Similarly, symmetric laminates had higher tensile strength compared to their asymmetric counterparts. Such behavior was mostly due to different fiber stackings that contributed to the crack deflection and their propagation.
3. Presence of adjacent plies with similar fiber orientation behaved like a single lamina, which in turn presented better crack-arresting properties.
4. Laminates in which adjacent plies angle difference was less had higher ultimate strength, depicting crack propagation is mostly endorsed by the difference between the orientation, where large difference causes sufficient damage and smaller one inhibits the same.
5. Similar behavior was obtained in laminates undergoing lateral loading since the stipulated reasons for crack initiation and propagation remained same, independent of loading pattern.

References

1. Ansari, M.T.A., Singh, K.K., Azam, M.S.: Fatigue damage analysis of fiber reinforced polymer composites—a review. J. Reinf. Plast. Compos. **37**(9), 636–654 (2018)
2. Dharan, C.K.H.: Fatigue failure in graphite fiber and glass fiber polymer composites. J. Mater. Sci. **10**(10), 1665–1670 (1975)
3. Gaurav, A., Singh, K.K.: Fatigue behavior of FRP composites and CNT-Embedded FRP composites: a review. Polym. Compos. **39**(6), 1785–1808 (2018)
4. Jones, R.M.: Mechanics of Composite Materials, 2nd edn. CRC Press Taylor & Francis, New York (1999)
5. Singh, K.K., Singh, R.K., Chandel, P.S., Kumar, P.: An asymmetric FRP laminate with a circular pre-crack to determine impact-induced damage. Polym. Compos. **29**(12), 1378–1383 (2008)
6. Gaurav, A., Singh, K.K.: ILSS improvement of quasi-isotropic glass fiber reinforced epoxy laminate enhanced with arc discharged multi-walled carbon nanotubes. Mater. Today Proc. **5**, 8638–8644 (2018)

7. Singh, R.K., Gaurav, A., Singh, K.K., Rizvi, M.R.: Effect of notch position and orientation on flexural behavior of neat epoxy GFRP laminates. In: IOP Conferences Series: Materials Science and Engineering, vol. 377, p. 012165 (2018)
8. Nunes, J.P., Pouzada, A.S., Bernardo, C.A.: The use of a three-point support flexural test to predict the stiffness of anisotropic composite plates in bending. Polym. Testing **21**(1), 27–33 (2002)
9. Ullah, H., Harland, A.R., Lucas, T., Price, D., Silberschmidt, V.V.: Finite-element modeling of bending of CFRP laminates: multiple delaminations. Comput. Mater. Sci. **52**(1), 147–156 (2012)
10. Rizvi, M.R., Singh, K.K., Gaurav, A., Singh, R.K.: Effect of strain rate on flexure properties of GFRP laminates—an experimental and numerical investigation. In: IOP Conferences Series: Materials Science and Engineering, vol. 377, p. 012085 (2018)
11. Dong, C., Davies, I.J.: Flexural strength of bidirectional hybrid epoxy composites reinforced by E glass and T700S carbon fibers. Compos. Part B **72**(1), 65–71 (2015)
12. Mini, K.M., Lakshmanan, M., Mathew, L., Mukundan, M.: Effect of fiber volume fraction on fatigue behavior of the glass fiber reinforced composite. Fatigue Fract. Eng. Mater. Struct. **35**(12), 1160 (2012)
13. Rehan, M.S.M., Rousse, J., Gonga, X.J., Guillaumat, L., Ali, J.S.M.: Effects of fiber orientation of adjacent plies on the mode I crack propagation in a carbon-epoxy laminate. Procedia Eng. **10**, 3179–3184 (2011)
14. Yokosuka, T., Aoki, T., Ogasawara, T., Ishikawa, T.: Effects of layup angle and ply thickness on matrix crack interaction in contiguous plies of composite laminates. Compos. Part A **36**(9), 1229–123 (2005)
15. Jacob, M., Thomas, S., Varughese, K.T.: Mechanical properties of sisal/oil palm hybrid fiber reinforced natural rubber composites. Compos. Sci. Technol. **64**(7–8), 955–965 (2004)

Chapter 7
Effect of Span-to-Depth Ratio on the Flexural Properties of Woven Neat Epoxy/Glass Fiber-Reinforced Polymer Symmetric Laminates

Punit Kumar Pandey, Kalyan Kumar Singh and Anand Gaurav

Abstract In this experimental study, the effect of span-to-depth ratio on flexural properties of woven glass fiber symmetric laminates was investigated. Two span-to-depth ratios of 15:1 and 30:1 were selected to assess the effect of span length on the flexure properties of woven GFRP laminates. Five samples from each span-to-depth ratio were tested according to the ASTM D790-17 standard under 3-point bending mode on Hounsfield H50KS, a computer-controlled universal testing machine. Tests revealed that flexure strength of woven GFRP laminates depends upon its span length. However, flexure strain to failure remained almost in the similar range for the specimen with different span lengths. Average flexural strength and modulus of samples with 15:1 and 30:1 span-to-depth ratio were obtained as 160.56 MPa, 14.99 GPa and 139.06 MPa, 19.17 GPa, respectively. Similarly, average strains to failure in the samples were found to be 0.01215 and 0.00853% in the respective order.

Keywords GFRP · Flexure · ASTM D790

7.1 Introduction

Due to merits like high specific strength (ratio of strength to the density of material), high specific modulus (Young's modulus per unit density of material), corrosion resistance, retaining strength at elevated temperature, etc., made polymer composites earn their applications in structural and engineering sectors like automobile, aerospace, construction, etc. [1]. High specific strength and high specific modulus are very noteworthy and consequential properties of composites because it exposits a quality of composites that is being strong and stiff along with light weightiness. In spite of having such advantages, high manufacturing cost, complex repairing process, time-consuming analysis and modeling of composites in comparison with

P. K. Pandey (✉) · K. K. Singh · A. Gaurav
Department of Mechanical Engineering, Indian Institute of Technology (Indian School of Mines), Dhanbad 826004, Jharkhand, India
e-mail: impunitpandey@gmail.com

I. Singh et al. (eds.), *Trends in Materials Engineering*,
Lecture Notes on Multidisciplinary Industrial Engineering,
https://doi.org/10.1007/978-981-13-9016-6_7

metals hinder their full-fledged implementation in engineering structure. However, their dominating advantages over disadvantages prompt expanding applications of composites.

Fiber-reinforced polymers (FRPs) are made up by matrix and high strength fibers. Characteristics of FRPs are amalgamation of its constituents. Main objective of the fibers is to bear the loads while matrix retains the fibers in required site and orientation and provides safeguard to the fibers against environment. Flexural properties analysis and study is crucial because material specimen experiences tension on one side of material, compression on other side of specimen and somewhere shear in between those two side's surfaces which gives mix mode of failure in the flexural test. So the investigation of effect of different factors on the flexural properties of composites is a major concern.

Epoxies are modern engineering polymers that have advantages like better toughness, moldability and low densities [2] compared to their other plastic counterparts. But they possess disadvantages like low fracture toughness and brittleness [3]. Their ease of handling and application on fibers to produce fiber-reinforced composites makes them a worthy choice for the fabrication of the same [2, 3]. Fiber-reinforced composites undergo several mechanical testing like tensile, compressive [4], impact [5] and fatigue [6] to assess their properties. A majority of these properties falls into fiber dominated one, while a few like flexure and ILSS are done to assess the matrix-dominated properties. Among the above-said two testings, in the previous one, sample undergoes a combination of tensile, compressive and shear failures. ASTM standard [7] for the stipulated testing considers correction factor, if the thickness-to-span ratio increases beyond 16:1.

In this research work, effect of span-to-depth ratio had been studied on woven epoxy glass-reinforced symmetric laminates. Sudarisman and Davies [8] evaluated the influence of span-to-depth ratio on unidirectional GFRP using 16:1, 32:1 and 64:1 span-to-depth ratios and found that flexural strength of the shorter beam was less than longer one. Dong et al. [9] checked the effect of span-to-depth ratio on glass and carbon hybrid composites and concluded that flexural modulus increases on increasing span-to-depth ratio from 16:1 to 32:1 but remained constant on further increase in the span length. Marcelo Rosensaft et al. [10] concluded that span-to-depth ratio can alter the relative magnitude of stresses.

But unidirectional laminates are only useful in in-plane loading condition such as uniaxial tension or uniaxial compression in fiber direction. They lack in transverse strength or through thickness strength and are often prone to delamination. Woven fiber laminates, on the other hand, have better in-depth properties compared to the previous one. So in this work, the effect of span length on the flexure properties of woven GFRP laminates is studied.

7.2 Materials, Fabrication and Processes

The material for experiment woven GFRP was developed by hand lay-up technique and cured in press molding machine. Reinforcement in the form of plain-weave glass fabric with surface weight of 610 GSM was selected. The material was manufactured by Vetrotex India Pvt. Ltd. and supplied by MS Industries, Kolkata, India. Room curing chemicals Lapox L-12 (ARL-12) brand's Bisphenol A-based thermosetting epoxy and N-N′-Bis (2-aminoethyl) ethane-1, 2-diamine hardener of Lapox K-6 (ARL-12) mixed in the ratio of 10:1 was used as matrix material. The chemicals were manufactured and supplied by Atul Industries, Valsad, Gujarat. Four laminae of (0°/90°) and four laminae of (+45°/−45°) were taken from woven glass fabric mat. For a bi-woven fabric, two numbers in the parenthesis show warp and weft direction of the fibers running through each ply. From these, 8-ply symmetric laminate was prepared which has the proposed stacking sequence of [(0°/90°), (+45°/−45°), (−45°/+45°), (90°/0°)]. Laminae were wetted with epoxy resin by a soft brush and stacked according to the proposed stacking sequence. After placing every lamina over previous one, laminate was rolled by mild steel roller to get rid of excess epoxy resin and expel trapped air between adjacent laminae. After that, a polythene bag was used for sealing purpose of laminate to get smooth surfaces after curing. Later, sealed laminate is placed under press molding machine at 25 KN for 24 h at room temperature to cure. Figure 7.1 shows the press molding machine and the prepared laminate.

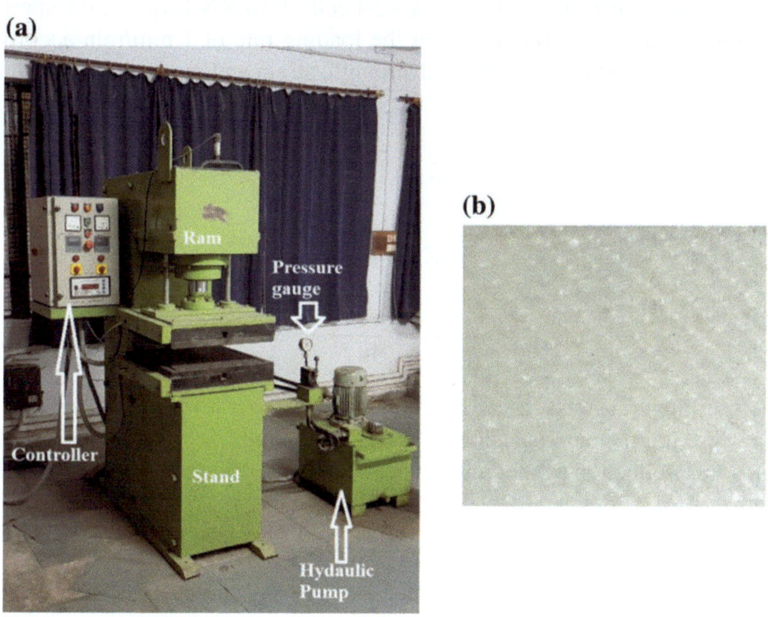

Fig. 7.1 Material fabrication method: **a** press molding machine and **b** prepared laminate

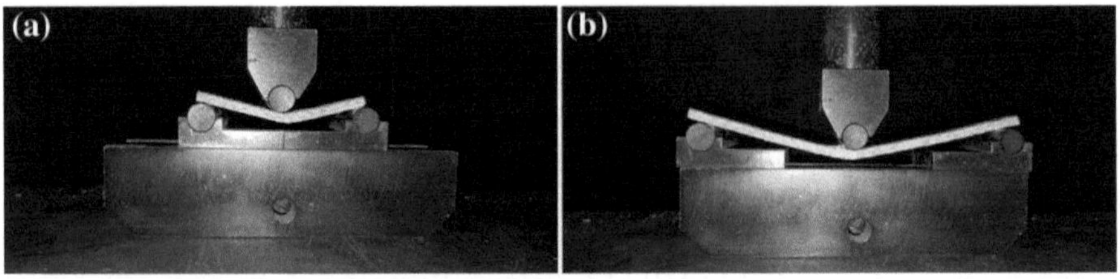

Fig. 7.2 Samples undergoing 3-PBT **a** 15:1 span-to-depth **b** 30:1 span-to-depth

After completion of curing process, laminate was taken out from press molding machine and rough edges were cut down to get proper laminate having squareness with edges. Then, the average thickness of laminate was measured using digital calipers and found to be 3.8 mm. After that as per the experiment, samples with 15:1 and 30:1 span-to-depth ratios were prepared from the laminate. After this, the edges were smoothened using emery paper to wear out burrs and irregularities. Five specimens for each span-to-depth ratio were prepared and tested.

7.3 Testing

Testing of the samples was carried out on a computer-controlled universal testing machine (Hounsfield H50KS) having a load cell of 50 KN (Fig. 7.2). Experiments were conducted using 3-PBT fixture at the loading rate of 1 mm/min according to the ASTM D790-17 standard. Flexural strength, strain and modulus were calculated as [7]:

$$\sigma = \frac{3PL}{2wt^2} \tag{7.1}$$

$$\mathcal{E} = \frac{6\delta t^2}{L^2} \tag{7.2}$$

$$E = \frac{mL^3}{4wt^3} \tag{7.3}$$

where σ = Flexural stress, $\mathcal{E}$ = Flexural strain, E = Flexural modulus.

P = Maximum load (peak load), L = Span length, w = width, t = thickness (depth), δ = Deflection of specimen at peak load , m = Initial slope of load–deflection curve.

Table 7.1 Flexure properties of the tested specimen with different span lengths with standard deviation

Span/depth	15:1			30:1		
Sample no.	σ (MPa)	E (Gpa)	ε	σ (Mpa)	E (Gpa)	ε
1	162.6	14.86	0.01206	128.1	18.37	0.00868
2	163.7	15.56	0.01347	150.0	19.54	0.00897
3	169.6	14.53	0.01203	136.9	19.45	0.00834
4	160.6	14.00	0.01191	134.1	19.41	0.00792
5	146.3	16.00	0.01128	146.2	19.09	0.00872
Average	*160.56*	*14.99*	*0.01215*	*139.06*	*19.172*	*0.00853*
S.D.	*8.64*	*0.80*	*0.0008*	*8.94*	*0.4793*	*0.0004*
COV (%)	*5.38*	*5.32*	*6.61*	*6.43*	*2.49*	*4.7*

7.4 Results and discussion

Table 7.1 presents the flexure properties of the tested specimen under 3-point loading. Standard deviation and coefficient of variance are presented to assess the randomness in the outcomes.

From the table, it can be observed that an increase in span-to-depth decreases flexure load-bearing capacity of the material and limits flexure strain to failure. On the contrary, an increase in flexural modulus is witnessed. Such aberration in the results regarding strength of the material could be due to the span length of the material. Samples with short span length fail by shearing in the mid-plane of the laminate, while a mix-mode fracture signifies the failure in samples with larger span length that causes global failure. It is believed that shorter beam fails by interlaminar shear forces, since upon loading matrix starts to deform more rapidly compared to the fibers.

Once the stresses in a loaded sample reach the shear strength of the matrix, sample fails in pure shearing without the load being transferred to the fibers and drop in the strength is observed. Once mid-plane matrix crack occurs, the fibers under the collective tensile and compressive load of the roller accumulate and provide additional reinforcement. But in the absence of matrix material, a collective fiber failure takes place known as fiber kinking [11]. These mechanisms collectively provide higher strain to failure for the samples with smaller span length (Fig. 7.3a).

On the other hand, specimens with larger span length witness global deformation throughout its length and transition from shear load to compressive/transverse load is observed [10]. Initial nonlinearity in the flexure stress–strain plot of the samples explains global failure and crack propagation in the material. Upon loading materials, internal defects (voids, resin fibrils, etc.) act as stress concentration and collectively hamper fiber–matrix interface. A sharp drop without the indication of mid-plane matrix failure (shearing) is observed due to fiber kinking and matrix shear-

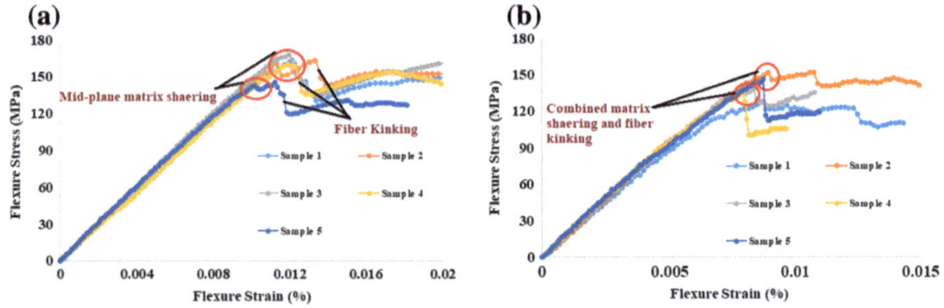

Fig. 7.3 Flexural stress versus flexural strain graph of samples: **a** 15:1 span-to-depth and **b** 30:1 span-to-depth

ing (Fig. 7.3b). This leads to lower strain to failure in the specimen with larger span length.

It was also observed that samples with larger span length depicted higher modulus compared to the one with smaller span length. In the earlier one, a sudden drop in the material strength at lower strain leads to higher stiffness, whereas in later one, matrix shearing that has already diminished the stiffness is followed by fiber kinking and material witness larger strain to failure (almost 1.5 times of the one presented by larger specimen). This leads to lower stiffness of the material with shorter span length. Similar results were presented by Dong et al. [9].

7.5 Conclusions

From the conducted tests, the following conclusion can be drawn.

1. On increasing span-to-depth ratio, flexural strength and failure strain decrease and flexural modulus increases.
2. Shorter samples failed by interlaminar shear at mid-plane while mix-mode fracture dictates failure in longer one.
3. Deviation in data of same batch of samples due to variation in flaws, misalignment and breaking of fibers affecting local tensile and compressive strength and ultimately flexural properties.
4. Shear damage occurs in the form of delamination just in the vicinity of loading roller.
5. Fraction of particular stress (tensile, compressive and shear stress) in total stress acting over specimen can be changed by changing span-to-depth ratio.
6. From design point of view, samples with large span length had lower strength. Hence, keeping factor of safety in mind larger span length is not recommended.

References

1. Gaurav, A., Singh, K.K.: Fatigue behavior of FRP composites and CNT-Embedded FRP composites: a review. Polym. Compos. **39**(6), 1785–1808 (2018)
2. Brydson, T.J.A.: Brydson's Plastic Materials (1966)
3. Hamouda, A.M.S., Risby, M.S.: Lightweight Ballistic Composites (2006)
4. Torabizadeh, M.A.: Tensile, compressive and shear properties of unidirectional glass/epoxy composites subjected to mechanical loading and low temperature services. Indian J. Eng. Mater. Sci. **20**, 299–309 (2013)
5. Cantwell, W.J., Morton, J.: the impact resistance of composites materials—a review. Composites **22**(5), 347–362 (1991)
6. Ansari, M.T.A., Singh, K.K., Azam, M.S.: Fatigue damage analysis of fiber-reinforced polymer composites-a review. J. Reinf. Plast. Compos. **37**(9), 636–654 (2018)
7. Standard test methods ASTM D790-17. Standard test methods for flexural properties of unreinforced and reinforced plastics and insulating materials
8. Sudarisman, Davies, I.J.: Flexural failure of unidirectional hybrid fibre–reinforced polymer (FRP) composites containing different grades of glass fibre. Adv. Mater. Res. **41–42**, 357–362 (2008)
9. Dong, C., Davies, I.J.: Optimal design for the flexural behaviour of glass and carbon fibre reinforced polymer hybrid composites. Mater. Des. **37**, 450–457 (2012)
10. Rosensaft, M., Marom, G.: Evaluation of bending test methods for composite materials. J. Compos. Technol. Res. **7**(l), 12–16 (1985)
11. Cox, B.N., Dadkhan, M.S., Morris, W.L., Flintoff J.G.: Failure mechanisms of 3D woven composites in tension, compression, and bending. Acta Metall. Mater. **42**(12), 3967–3984 (1994)

Chapter 8
Dehumidification of Maize in Desiccant Dryer Using Thermal Imaging

Lakhan Agarwal and Ashok Yadav

Abstract Using a desiccant-based dryer, the drying chamber is tested in the heat transfer laboratory, Faculty of Engineering, Dayalbagh educational institute, Agra. Experiments are carried out to analyze the effect of heat on the drying maize at a different time and different drying trays positions. After harvesting, drying is an indispensable process for extraction of moisture content of grain production and improving overall performance for better outcomes. In India, sun drying is the main drying method used by farmers for drying their crop grain yield. The desiccant dryer is used for dehumidification of maize to remove moisture contents. The present research work emphasizes to analyze dehumidification on maize in the drying chamber at different time interval by using a thermal camera. A thermal camera is used to find the drying rate and performance evaluation of maize. Performance of dehumidifier is also correlated with the heat parameters. Thermal imaging gives a realistic and deeper view of grain drying. The efficiency was calculated to be 76.45% using desiccant drying.

Keywords Desiccant · Dehumidification · Heat · Dryer · Thermal image

8.1 Introduction

India is an agrarian economy with its 58% rural population engaged in agricultural activities placing it on the sixth rank globally in the food and grocery markets, with the retail industry contributing to a humungous 70% of total sales. The condition of the Indian farmer is poor due to lack of mean of agriculture techniques. The Large volume of grains is produced by Indian farmer, but due to the lack of drying and storage capabilities, farmer does not get proper benefits. Traditionally, Indian

L. Agarwal · A. Yadav (✉)
Mechanical Engineering Department, Dayalbagh Educational Institute, Agra 282005, India
e-mail: ashokyadavaca@gmail.com

L. Agarwal
e-mail: lakhanagarwal24@gmail.com

I. Singh et al. (eds.), *Trends in Materials Engineering*,
Lecture Notes on Multidisciplinary Industrial Engineering,
https://doi.org/10.1007/978-981-13-9016-6_8

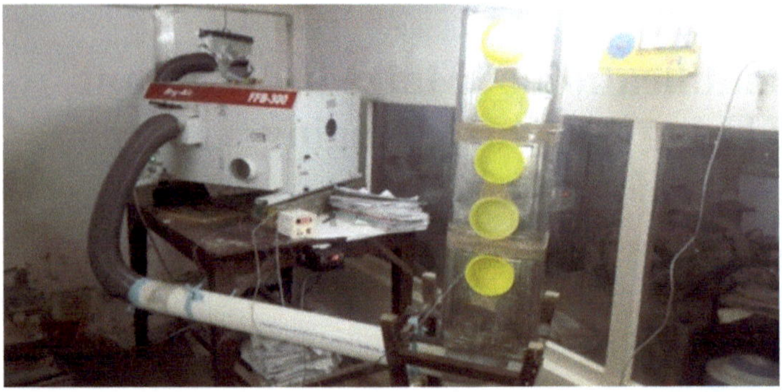

Fig. 8.1 Desiccant dehumidifier system

farmer use sun drying for various grains, which is dependent on weather condition, and fluctuation of heat and temperature occur several times. Also, sun drying is not hygienic and does not have control over drying.

In general, drying is accomplished by thermal techniques and thus involves the application of heat. Throughout drying two processes occurs namely heat transfer as well as mass transfer. Many designs have been proposed for forced convection solar dryer to achieve proper drying rates [1]. A simple effective solar dryer based on desiccant drying will use solar radiation as heat requirement for charging the desiccant material. It will be simple to use effective in performance.

8.2 Desiccant Dehumidification System

Desiccant Materials have a property to adsorb moisture conditioned to the differences in the vapor pressure indices and due to the hankering propensity of the material to retain water content, it has found itself utility in a smorgasbord of applied industrial uses like pharmaceutical, chemicals, electronics, etc. [2]. As well as desiccant systems can create low-temperature drying at a controlled rate which is very important for certain agricultural products. Desiccants can eliminate moisture from the air which can be used for drying purposes. The weak desiccant can be regenerated by air under assured conditions so that the desiccant can be used continuously to remove moisture from the air. As hot and moistened air passes through the desiccant wheel, it encounters a gradual spur in the temperature as the moisture is removed. The applications of combining the resources from solar energy and desiccant energy storage [3]. Various physical and thermal properties have been complied. Desiccant which is found in both, liquid and solid forms, having pervasive pragmatic utilities is usable in both, small scale and large scale drying industries because of its anti-hydrophobic nature [4]. A dehumidifier model was developed which depicts a liaison between the outlet and inlet states of the dehumidifier (Fig 8.1).

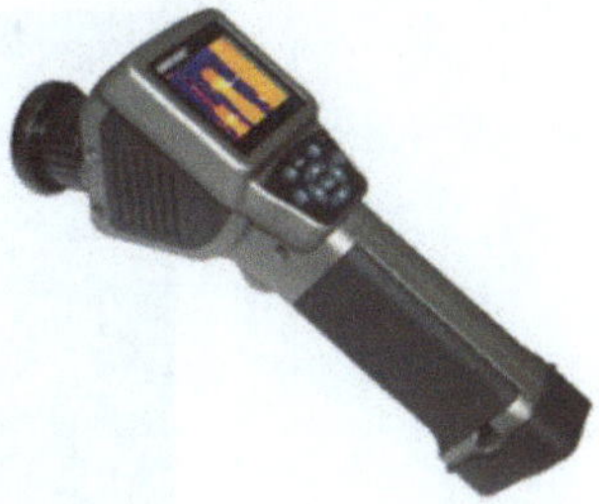

Fig. 8.2 KusamMecoThermal Imaging Camera

The heat gained by the chamber can be calculated as:

$$Q = mc(T - Ta)$$

where Q = heat gained/received by dehumidifier
m = mass flow rate
c = specific heat of (chamber/ machine outlet)
T = temperature (chamber/ machine outlet)
T_Q = Ambient temperature.
In general, mass flow rate can be calculated as:

$$m = \rho AV$$

ρ = density of heated air
A = area of duct
V = velocity of air [5].

8.3 Introduction to Thermal Imaging

Thermal imaging is a technique to detect thermal radiation above absolute zero temperature. Thermal imaging camera is a device used to pick up this infrared radiation which captures image known as thermogram or thermal image [6]. The thermal camera shows a colorful pattern of thermal images on the screen with a temperature scale mentioning temperature range of the captured image. The hot areas of temperature show dark color while colder areas show bright color (Fig. 8.2).

8.3.1 Temperature Measurement by Thermal Imaging

The moisture content in fresh agricultural products is the basic cause of spoilage. If moisture is removed from microorganisms, it will inhibit its growth and reproduction, neutralized moisture deteriorative reactions [7]. If water is removed, then the shelf

Fig. 8.3 Position of maize

life of the agricultural product can be increased. Nowadays various types of drying technologies are used for drying. Sun drying is a traditional method of preservation of grains. The quality of sun-dried product is poor with regard to its color and flavor due to direct exposure to sun [8]. If the temperature of the hot dry air passing through the drying chamber is in access,the quality, color, texture, and germination will be caused, which finally lead the damage and loss of grains. Normal visualization or with the help of thermostat of grains through naked eyes does not identify the heating. Thermal imaging is a good technique for measuring the surface temperature of the grain.

8.4 Dehumidification of Maize Using Thermal Imaging

A desiccant-based dryer, drying chamber tested in the heat transfer lab, faculty of engineering, Dayalbagh educational institute, Agra. The observations were taken on a rainy day and the wet maize is placed on the aluminum grilled tray in the drying chamber of the desiccant dryer. The initial conditions for experimental setup are mention below:

Grain type: Maize (wet for 3 h in water)
Room temperature: 32.4 °C
Air velocity of dehumidifier: 10 m/s.

The Figs. 8.3, 8.4, 8.5 and 8.6 show the maize to use for dehumidification, maize drying chamber with a humidistat sensor, a digital screen showing relative humidity of the maize drying chamber and anemometer. Figures 8.7, 8.8, 8.9 and 8.10 show thermal images of maize and maize drying chamber during dehumidification.

The following Table (8.1) shows the dehumidification parameters for maize at a different time interval of 1 min, at the same time relative of the chamber is recorded by humidistat by digital screen and chamber temperature is recorded by anemometer which is placed inside the chamber.

From Table (8.2), the specific heat for the experimental work is considered as 46.81–46.84 for the temp range between 35 and 41 for chamber and 46.82–46.87

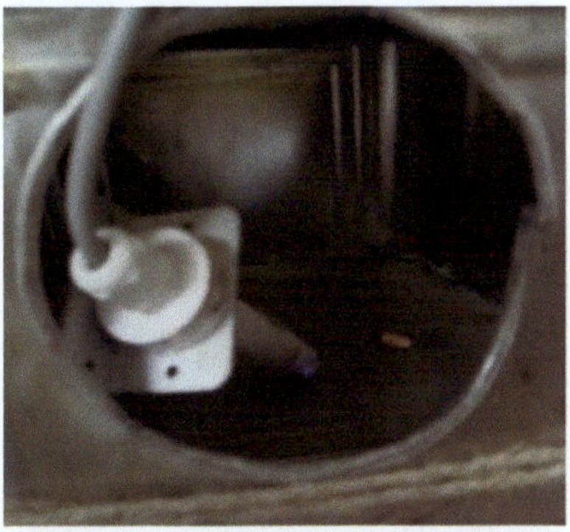

Fig. 8.4 Relative humidity checked by humidistat sensor

Fig. 8.5 Relative humidity on screen

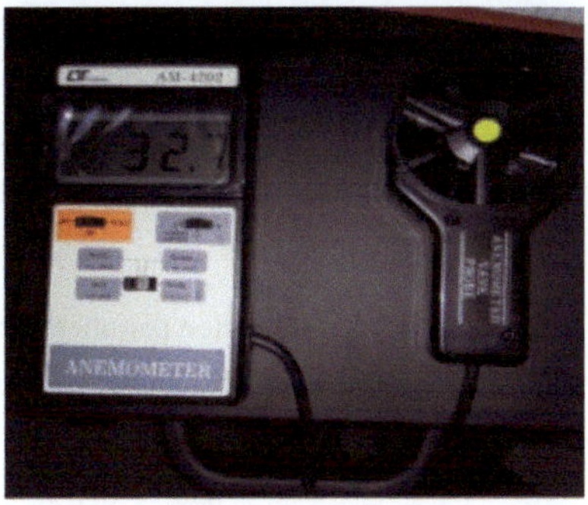

Fig. 8.6 Temperature measuring instrument (ANEMOMETER)

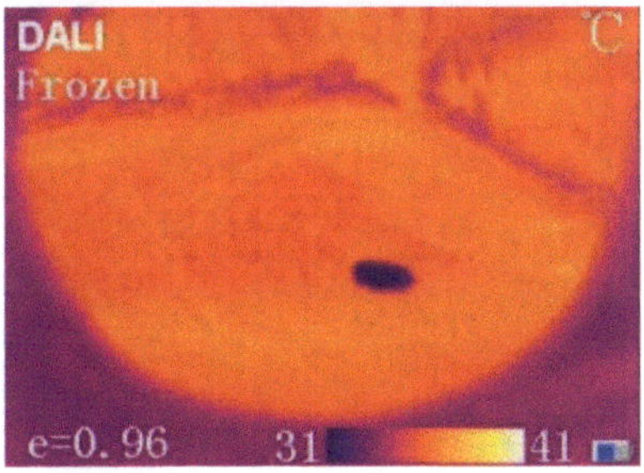

Fig. 8.7 Dark color of maize

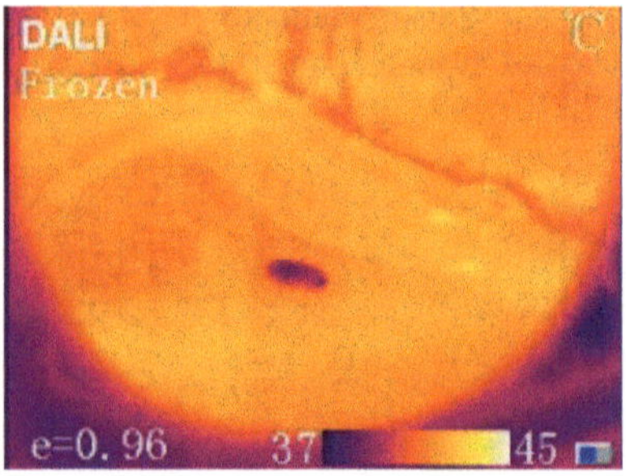

Fig. 8.8 Color lighter than Fig. 8.7

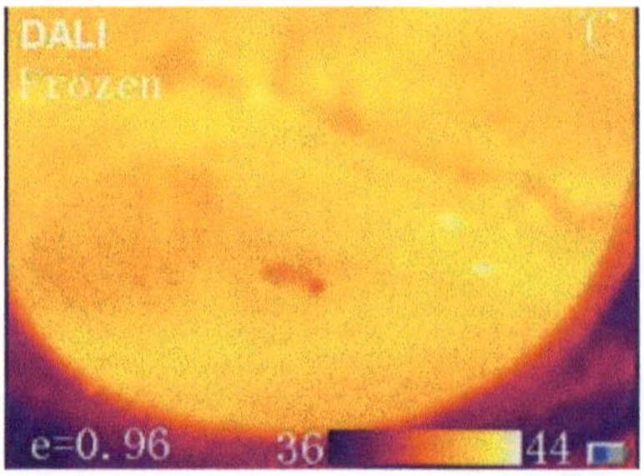

Fig. 8.9 Maize little bit seen than Fig. 8.8

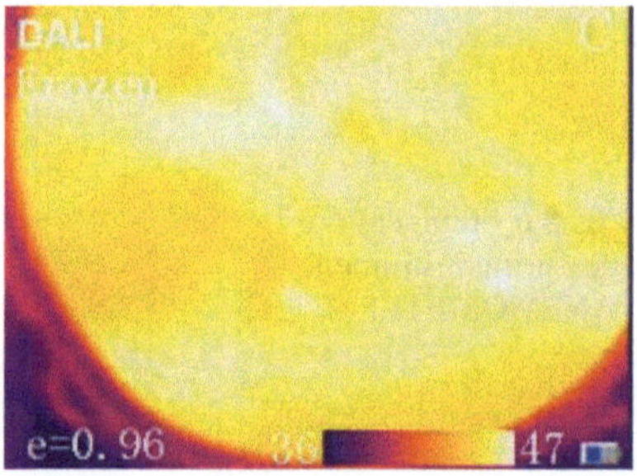

Fig. 8.10 Maize not seen clearly

Table 8.1 Dehumidification parameters for maize

S. No.	Chamber temperature (°C)	Chamber outlet specific heat C_o(kcal/kg)	Heat gained chamber(Q_O)	Machine temperature (°C)	Inlet specific heat C_i(kcal/kg)	Heat received(Q_i)
1	35	46.81	36.80	36	46.82	50.9
2	38	46.84	79.32	40	46.84	107
3	40	46.84	107.64	43	46.84	150.1
4	41	46.84	121.81	44	46.87	164

Table 8.2 Heat gained for chamber

S. No.	Date	Time (A.M)	Chamber relative humidity (%)	No. of grains (Maize)	Chamber temperature (°C)
1	27.08.2018	11:25	98	1	35
2		11:26	56		38
3		11:27	49		40
4		11:28	42		41

for the temp range between 36 and 44 for a dehumidifier. The chamber temperature varies from 35 to 41 °C (small range); therefore, the mass flow rate is considered as 0.3024 kg/h.

Average output heat gained by the chamber is calculated by taking mean of individual heat gained by the chamber:

$$Q_{oav} = 86.39$$

Similarly, the average output heat received by the machine is calculated by taking mean of individual heat received by the machine:

$$Q_{iav} = 113$$

The average efficiency of the desiccant dehumidifier system for maize is calculated as:

$$\eta = \frac{Q_{oav}}{Q_{iav}} \times 100 = 76.45\%$$

8.5 Result and Discussion

The work done focuses on the dehumidification analysis of maize grain with thermal imaging. The drying rate of maize becomes faster as the temperature of the drying chamber increases rapidly with the decrement in relative humidity of the drying chamber. This dehumidifier extracts the moisture content from the atmospheric air and this heated and dehumidified air is used to dry the maize to achieve the process of dehumidification. The thermal images captured by a thermal imaging camera show the temperature of maize and drying chamber. Figures 8.7, 8.8, 8.9 and 8.10 show the different stages obtained for the maize placed in the drying chamber during dehumidification with dark color representing low temperature and light color representing high temperature. This color variation can be easily seen in the thermal images of the maize with a drying chamber. The variation range between minimum

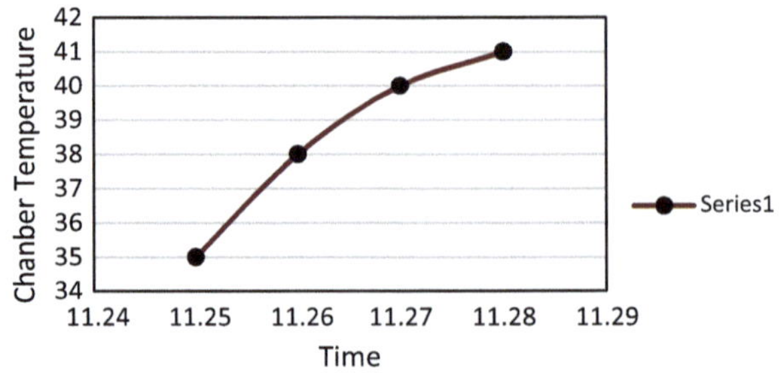

Fig. 8.11 Variation of chamber temperature (°C)

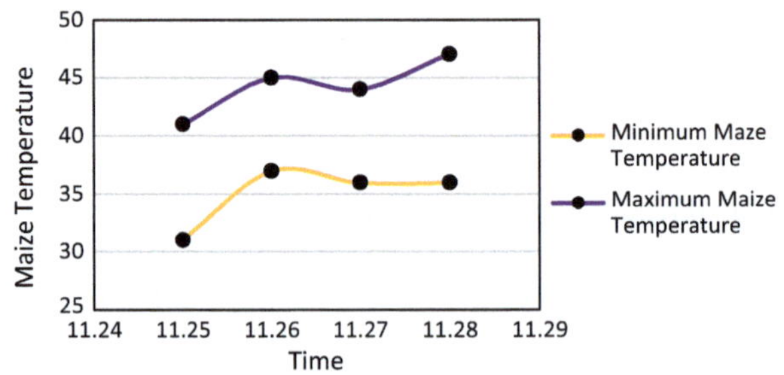

Fig. 8.12 Variations in maize temperature (°C)

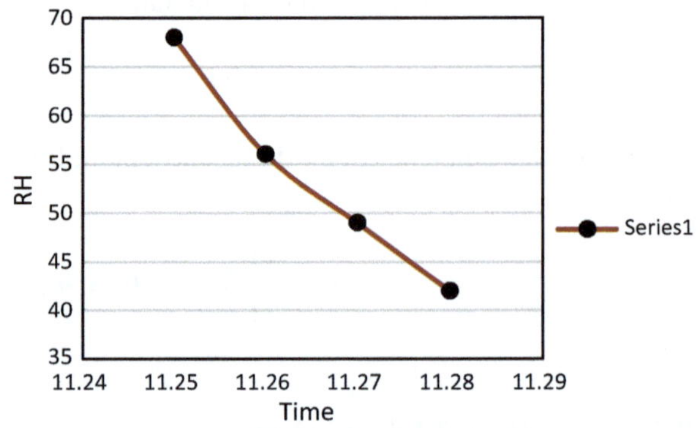

Fig. 8.13 Variation of relative humidity in drying chamber (%)

and maximum temperature of wet maize grain to be dry are shown in Figs. 8.11, 8.12 and 8.13 shows an inverse relation between relative humidity and maize drying temperature.

The experimental work done in this paper concludes a study on the drying time of the maize for a very short span of time, i.e., for three minutes which is also shown by the graphs obtained from the observation table during the dehumidification process. Indian farmer, using traditional sun drying process for various grains, is dependent on weather condition. In this process the fluctuation of heat and temperature occurs several times. Also, sun drying is not hygienic and does not have control over drying. Using this type of desiccant dryer system, the above problems will be solved as the rate of drying of grain is uniform throughout the process.

References

1. Mohanraj, M., Chandrasekar, P.: Performance of a forced convection solar drier integrated with gravel as heat storage material for chillidrying. J. Eng. Sci. Technol. **4**(3), 305–314 (2009)
2. Wurm, J., Kosar, D., Clemens, T.: Solid desiccant technology review. Bull. Int. Inst. Refrig. **82**(3), 2–31 (2002)
3. Miller, W.M.: Energy storage via desiccant for food agricultural applications. Energy Agri. **2**, 341–354 (1983)
4. Smith, R.R., Hwang, C.: Modelling of a solar-assisted desiccant air conditioner for a residential building, ^{1}hermodynamics and the Design. Anal. Improv. Energy Syst. **266**, 409–418 (1992)
5. Seveda, M.S., Jhajharia, D.: Design and performance evaluation of solar dryer for drying of large cardamom(Amomumsubulatum). J. Renew. Sustain. Energy **4**(6), 1–11 (2012)
6. Chaudhary, A.S., Chaturvedi, D.K.: Temperature effects bird and cement deposits on solar panels, pp. 86–92. Electrical India, Chary Publications Mumbai (2018)
7. Dincer, M., Hussain, M., Sahin, A.Z., Yilbas, B.S.: Development of a new moisture transfer (Bi–Re) correlation for food drying applications. Int. J. Heat Mass Transfer **45**, 1749–1755 (2002)
8. Jayaraman, K.S., Das Gupta, D.K., Rao, N.: Quality characteristics of some vegetables dried by direct and indirect sun drying. Ind Food Packer **45**(1), 16–23 (1991)

Chapter 9
Influence of Oblique Impact on Glass Fiber-Reinforced Polymer Composites: A Numerical Approach

Prashant Rawat, Nand Kishore Singh, K. K. Singh and Nitesh Agrhari

Abstract This paper investigates the low-velocity impact response of a glass fiber-reinforced polymer (GFRP) composite. A four-layered anisotropic bidirectional glass fiber composite laminate is modeled using HyperMesh. The thickness of each ply is considered as 0.5 mm, and the radius of the impactor is 10 mm. The oblique angle impact at 0°, 15°, 30°, and 45° (inclination of laminates with the horizontal plane) is simulated using LS-DYNA, which is a 3D commercially available software. A comparative study is carried out for energy absorption, stress generation, damage caused, and deflection of the laminate at 3 m/s impact velocity. The results of the mechanical testing simulations justified that the impact angle highly influences the impact response. Therefore, it may help in designing mechanical components that are exposed to oblique low-velocity impact (LVI) situations.

Keywords LVI · GFRP · Angular impact · Energy absorption · Damage area

9.1 Introduction

Laminated polymer composites have been frequently used as a low-weight, high-specific stress, and stiffness properties material. The application of fiber/epoxy composites is in automobiles, aerospace, marine industries due to extraordinary tailored properties. However, these materials are also susceptible to damage under impact

P. Rawat (✉)
College of Civil Engineering, Hunan University, Changsha 410082, Hunan, P.R. China
e-mail: aadiprashant@gmail.com

N. K. Singh
Mechanical Engineering, Dayananda Sagar University, Bangalore 560078, India

P. Rawat · K. K. Singh
Department of Mechanical Engineering, Indian Institute of Technology (ISM), Dhanbad 826004, India

N. Agrhari
Mechaniacal Engineering, Delhi Technical Campus, Greater Nodia 201306, India

I. Singh et al. (eds.), *Trends in Materials Engineering*,
Lecture Notes on Multidisciplinary Industrial Engineering,
https://doi.org/10.1007/978-981-13-9016-6_9

loading [1]. These damages are completely different as compared to metallic components having isotropic properties. In fiber reinforced polymer (FRP) composites the damage occurred in several modes like crack initiation; crack branching, delamination and finally fiber fracture [2]. The damage caused also influenced by impactor position and the modes of damage may be delamination at different interfaces, subsurface failure [3]. Therefore, it becomes necessary to investigate the damage caused at different oblique angle situations, as this kind of damage is mostly barely visible impact (BVI) in nature. Also, they degrade the strength of the material at faster rate which is not a favorable situation [4].

Madjidi et al. [5] in 1996 investigated the oblique impact of Chopped Strand Mat based polymer composites at 59 J impact energy at 0–300 angles plate angle variations. This research highlighted the importance of oblique impact as they concluded that the matrix cracking developed was diagonally through the thickness depending on the angle of inclination. Fadzullah et al. [6] investigated the penetration behavior of PET foam at 100 and 200 angular position and 3.78 m/s impact velocity. They highlighted the damage area which depends upon the inclination angles and increases with increment in inclination.

The main purpose of the present work is to numerically analyze the effect of oblique impact on FRP laminates made of bidirectional weaved fabrics. This kind of simulation is used to predict energy absorbed, stress developed, and damage generated in composite laminates made of bidirectional fabric under oblique low-velocity impact. In addition, the components, which are exposed to LVI situations, can be designed in such a way that the minimum risk of damage is present. This is the novelty of the presented paper.

9.2 Problem Description

Previous work [7, 8] was performed to analyze the basic aspect of transverse impact and pre-cracked samples with different layups. However, no oblique angular position was considered, while it also influences the LVI damage performance in a significant manner. In order to overcome this limitation, a four-layered bidirectional glass woven composite laminate (with 10 cm $\times$ 10 cm $\times$ 2 mm) is modeled in HyperMesh V9.0 with 1 mm^2 element size. This model is imported in LS-DYNA 4.2 (beta) for low-velocity impact simulations at 3 m/s impact velocity. The inclination of this laminate with the horizontal plane is 0°, 15°, 30°, and 45° for low-velocity impact in transverse direction as shown in Fig. 9.1. Five simulations for each case are performed in order to get the best results.

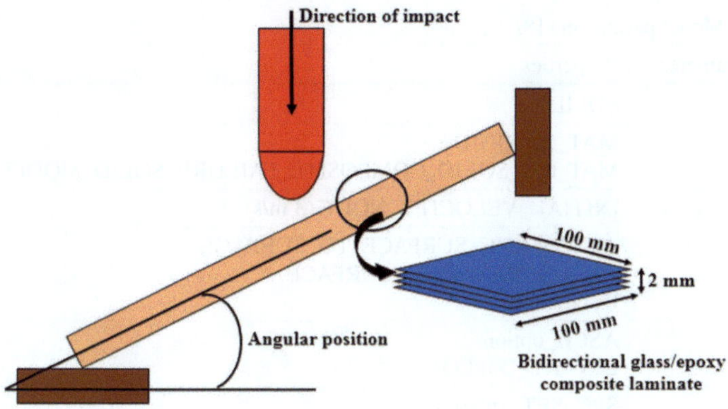

Fig. 9.1 Schematic illustration of impact events

9.3 Materials and Its Properties

The glass woven of 600 GSM is used as a reinforcement while Bisphenol A and K-6 hardener (supplied by Atul Ltd, Gujrat, India) mixed as a matrix material. The ratio of epoxy and hardener mixing is 10:1 (by weight) as suggested by the supplier. Composite laminate is fabricated by vacuum bagging method; the mechanical properties are calculated for neat GFRP laminates [9]. These properties are used for the simulating oblique impact tests and summarized in Table 9.1.

Table 9.1 Material properties [9]

GFRP bidirectional laminate		Properties
Young's modulus (GPa)	E_{11}	26
	E_{22}	26
	E_{33}	8
Poisson's ratio	v_{21}	0.1
	V_{31}	0.25
	v_{32}	0.25
Tensile strength (GPa)	T_x	0.850
	T_y	0.850
	T_z	0.120
Compressive strength (GPa)	C_x	0.720
	C_y	0.720
	C_z	0.5
Shear strength (GPa)	S_{12}	0.105
	S_{13}	0.065
	S_{23}	0.065

Table 9.2 Model parameters [8]

Model parameter	Properties
Section	SOLID
Materials	MAT_20_RIGID MAT_059_SOLID_COMPOSITE_FAILURE_SOLID_MODEL
Velocity	INITIAL_VELOCITY_NODE (3 m/s)
Contact	AUTOMATIC_SURFACE_TO_SURFACE TIED_SURFACE_TO_SURFACE INTERIOR
Database	ASCII_option BINARY_D3PLOT
Boundary	SPC_SET (sphe)
Control	ENERGY TERMINATION (3 ms)

The modeling meshing of the bidirectional lamina is done using HyperMesh with the element size of 1 mm^2. Four-layered laminate (bidirectional with 0°/90° ply orientations) and an impactor are modeled and imported to LS-DYNA to define the material properties and boundary conditions. The various model parameters are summarized in Table 9.2.

MAT_59_Composite_Failure_Solid_Model is a material model for orthotropic material and works on maximum failure criterion for mechanical loading. Failure due to tensile, compressive, shear, and delamination is covered in this material card. Simulation of low-velocity impact (using MAT_59) on four-layered GFRP laminate consists of the following failure mechanism (material models): longitudinal tensile, transverse tensile failure, through-thickness shear failure (longitudinal), through-thickness shear failure (transverse), delamination failure (through-thickness tension), longitudinal compressive failure, transverse compressive failure, through-thickness compressive failure.

9.4 Results and Discussion

9.4.1 Energy Absorption

The energy absorption by the laminate is compared based on laminate's internal energy, and it varied with the angular placement of the composite plate. The plots shown in Fig. 9.2, the internal energy is minimum, i.e., 8.1 J for normal horizontal laminate. This energy is minimum as complete fracture took place and the laminate releases energy. While after reaching, maximum value of energy various peaks is observed that shows the layer-by-layer fracture of the composite laminate.

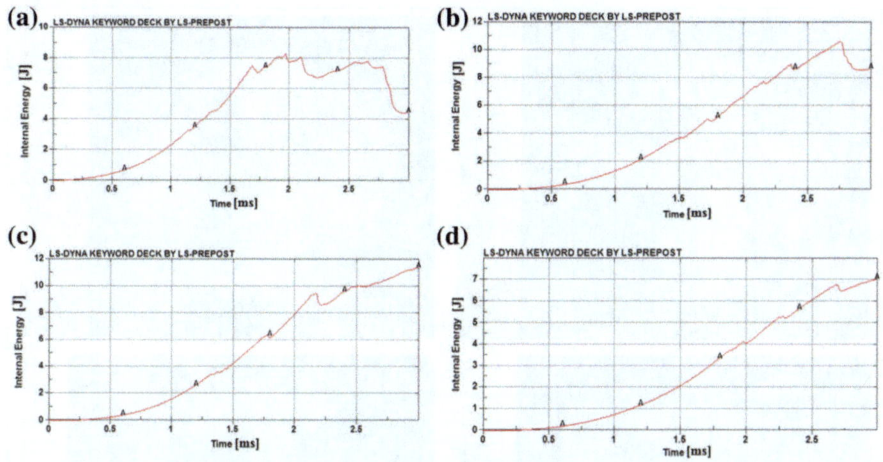

Fig. 9.2 Internal energy versus time plots **a** for the impact angle of 0°, **b** for the impact angle of 15°, **c** for the impact angle of 30°, and **d** for the impact angle of 45°

In Fig. 9.2b, the time to reach peak energy is higher as the laminate is inclined at an angle of 15°; this situation causes the delamination (subsurface failure) as a horizontal component of impact velocity acted as a shear force over the laminate. This shear loading helps increase in internal energy, and causes multilayer delamination [3] while at maximum angle i.e. at 45° the energy absorption is minimum that implies that with the increase in inclination angle of impactor the energy absorption increased up certain angle and then start decreasing. Moreover, the oblique impact situation influences on damage modes as well as internal energy of the laminate.

9.4.2 Stress Generation

The stress pattern shown in Fig. 9.3a–d highlights the peak stress generated during the impact. The stress pattern is different in every case which indicates that the modes of damage are varying with the variation in angular positions of the impact. Lowest value of maximum stress is in horizontal plate and transverse impact condition. The reason for this minimum value is that the stress is released with the fracture or complete penetration. Highest stress is generated for 15° angular position, the cause of this highest value is no ply failure, and minimum delamination is caused; therefore, the stress generation is maximum.

While in Fig. 9.3c the stress decreases highlighting the maximum subsurface failure but no sign of complete ply fracture. Whereas at angular position of 45° it started increasing this implies the subsurface failures are less due to the high angle of inclination. This comparison justifies an important aspect of material response at oblique impact situation.

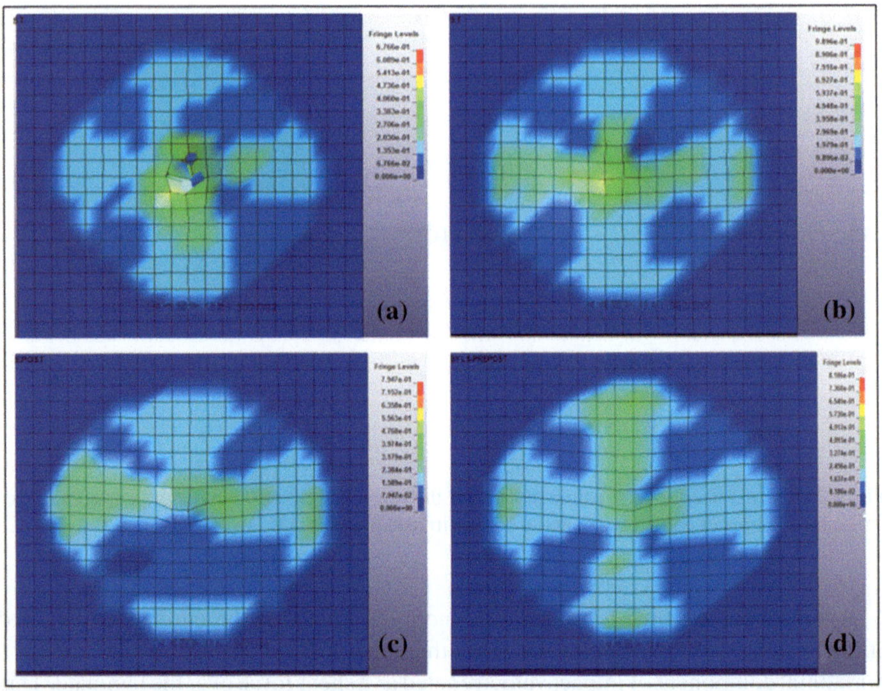

Fig. 9.3 Stress pattern at different position of ply **a** at 0°, **b** at 15°, **c** at 30°, and **d** at 45° oblique angles

9.4.3 Damage Area

Maximum rectangular damage area and pyramidal damages are compared and are shown in Figs. 9.4 and 9.5. The damage caused in fiber/epoxy composites is highly dependent on modes of failure. At 0° inclined specimen, the major part of damage is penetration of impactor inside laminate based, although for other oblique angular situations multilayer failure damage and distorted swelling of laminates are observed. Therefore, the damage area is compared on the basis of top and side view.

In Fig. 9.4a (top view), the rectangular area measured is 29.80 × 24.88 mm^2; in Fig. 9.4b, the damage area is 29.66 × 24.79 mm^2. However, the major difference in both cases is complete penetration or fiber fracture at 0° incident angle, whereas at 15° angle no penetration took place as shown in Fig. 9.5a–b. With the increase in oblique angle, these modes of fracture played a significant role, and at 30° inclination angle, the damage area is 24.51 × 19.73 mm^2. Whereas at highest angle of inclination, i.e., at 45°, the rectangular damage area is observed negligible as compared to other cases, only delamination of ply took place, and only swelling of the laminate is detected for this case (Fig. 9.5d).

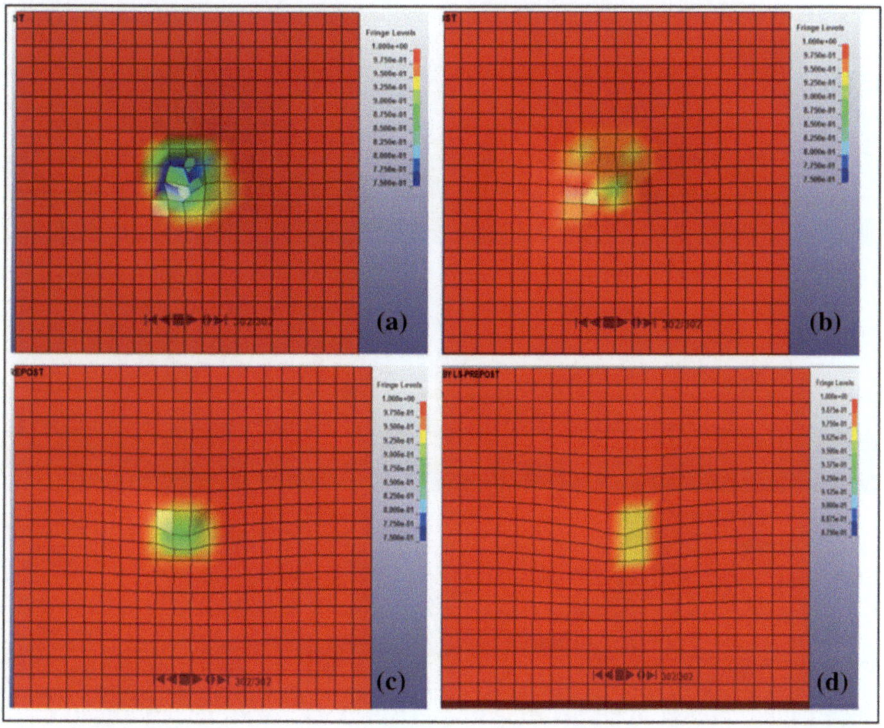

Fig. 9.4 Damage area of low-velocity impact **a** at 0°, **b** at 15°, **c** at 30°, and **d** at 45° oblique angles

9.4.4 Deflection of the Laminate

With the variation in oblique angles, at same incident velocity different kinds of damage pattern are observed. Therefore, it is also necessary to understand and compare the deflection/deformation of the laminate at various oblique angles. Figure 9.6a–d shows the displacement of the laminate in *Y*-direction. Here, the layer-by-layer displacement is plotted with respect to time.

Figure 9.6a shows the displacement of each layer follow different pattern, as the complete fracture occurred for this case, the top most layer exposed to the impactor showed positive displacement. The reason for this behavior is the loading of sample under impact situation; i.e., multiple loading tensile and compression take place on both the surface top and bottom surfaces [10]. Due to this multiple loading at the top surface of the laminate, some swelling is observed, i.e., curve A2 in Fig. 9.6a, whereas rest layers showed only negative displacement but different nature. This is due to the layer-by-layer fracture of the laminate.

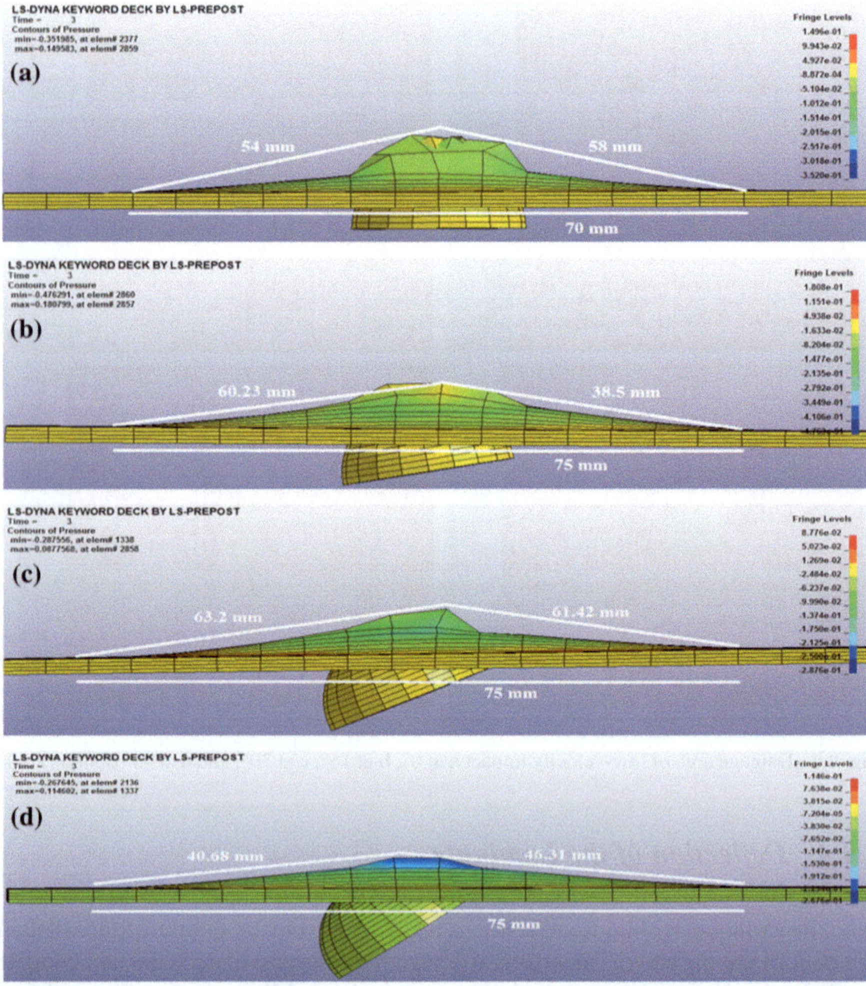

Fig. 9.5 Damage area (front view) of low-velocity impact **a** at 0°, **b** at 15°, **c** at 30°, and **d** at 45° oblique angles

Figure 9.6b–c did not show any major characteristic in displacement curves, and the main cause of this phenomenon was the only subsurface failure. In addition, it is also clear that during the subsurface failure, the shear loading plays a major role, and this load is the due to the horizontal component of the applied load, whereas vertical component causes swelling of the laminate only.

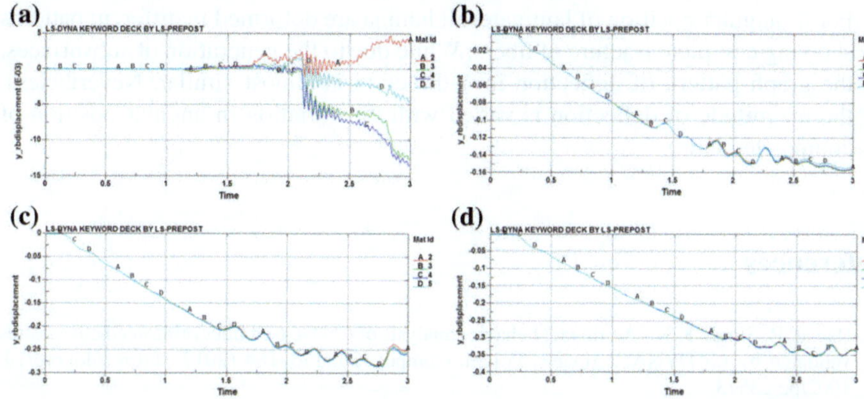

Fig. 9.6 Displacement (in mm) versus time (in ms) curves **a** at 0°, **b** at 15°, **c** at 30°, and **d** at 45° oblique angles

9.5 Conclusion

A numerical simulation-based investigation has been done to understand the influence of different oblique angle over GFRP laminate under low-velocity impact. The comparison of impact damage in terms of energy absorption, stress pattern, damage, and laminate deflection has been carried out. The outcome of this research can be listed as:

1. The energy absorption by the laminate is increased with the inclination angle while above certain angle the energy absorption start decreasing. The main cause of this decrement is the generation of random subsurfaces in between plies.
2. The stress or response of laminate under LVI is also influenced by oblique angle of laminate. For horizontally placed sample, it is uniformed and balanced about the center planes. However for angles of 15°, 30°, and 45° angular positions, the damage starts shifting from the center position depending upon the generation of delamination or fracture caused in between the plies.
3. Damage area is the most important factor to be considered as complete fracture causes visible damage while BVD due to low-velocity impact is most critical damage conditions to be considered and analyzed. In this paper, complete failure is detected for 0° angular placement of laminate. In addition, the rectangular damage area is nearly same for 0° and 15° inclinations, but no fiber failure is observed above in 15° oblique angle position. This justified that oblique indentation problem causes BVD in FRPs and needs to be considered as prime objective for the safety of the composite components.

4. For 0° angular position of laminate, all lamina are deformed in different patterns showing complete fracture of fibers. While due to the generation of subsurfaces, the graph pattern of deflection in Y-direction is almost similar. Nevertheless, the magnitude of deflection is varied with the variation in angular position of composite plate.

References

1. Rawat, P., Singh, K.K.: An impact behavior analysis of CNT-based fiber reinforced composites validated by LS-DYNA: a review. Polym. Compos. **38**, 175–184 (2017). https://doi.org/10.1002/pc.23573
2. Singh, K.K., Rawat, P.: Mechanical behavior of glass/epoxy composite laminate with varying amount of MWCNTs under different loadings. Mater. Res. Express. **5**, 55012 (2018). https://doi.org/10.1088/2053-1591/aabf99
3. Dey, S., Karmakar, A.: Effect of oblique angle on low velocity impact response of delaminated composite conical shells. Proc. Inst. Mech. Eng. Part C J. Mech. Eng. Sci. **228**, 2663–2677. https://doi.org/10.1177/0954406214521799
4. Richardson, M.O.W., Wisheart, M.J.: Review of low-velocity impact properties of composite materials. Compos. Part A Appl. Sci. Manuf. **27**, 1123–1131 (1996). https://doi.org/10.1016/1359-835X(96)00074-7
5. Madjidi, S., Arnold, W.S., Marshall, I.H.: Damage tolerance of CSM laminates subject to low velocity oblique impacts. Compos. Struct. **34**, 101–116 (1996). https://doi.org/10.1016/0263-8223(95)00137-9
6. Fadzullah, S.H.S.M., Ifayefunmi, O., Mustafa, Z., Omar, G., Mansor, M.R.: Perforation behaviour of composites sandwich structures on low-velocity impact at oblique angles. Int. J. Automot. Mech. Eng. **14**, 4158–4170 (2017). https://doi.org/10.15282/ijame.14.2.2017.4.0333
7. Rawat, P., Singh, K.K., Singh, N.K.: Numerical investigation of damage area due to different shape of impactors at low velocity impact of GFRP laminate. Mater. Today Proc. https://doi.org/10.1016/j.matpr.2017.07.222
8. Rawat, P., Singh, K.K., Singh, N.K.: Numerical investigation of low-velocity impact in symmetric and asymmetric GFRP laminate with and without pre-crack. Adv. Mater. Proc. **2**, 152–155 (2017). https://doi.org/10.5185/amp.2017/304
9. Singh, K.K., Singh, N.K., Jha, R.: Analysis of symmetric and asymmetric glass fiber reinforced plastic laminates subjected to low-velocity impact. J. Compos. Mater. **50**, 1853–1863 (2016). https://doi.org/10.1177/0021998315596594
10. Singh, K.K., Rawat, P.: Mechanical behavior of glass/epoxy composite laminate with varying amount of MWCNTs under different loadings. Mater. Res. Express. **5**. https://doi.org/10.1088/2053-1591/aabf99

Chapter 10
Numerical Simulation of GFRP Laminate Under Low-Velocity Impact at Different Edge-Constrained Boundary Conditions

Mahesh and K. K. Singh

Abstract FRP laminates are highly vulnerable to low-velocity impact (LVI) because it induces barely visible impact damage (BVID) inside the structure. This kind of fracture or damages is dangerous to the structure because these damages may go unnoticed ultimately leading to sudden and catastrophic failure of the structure. In this numerical simulation LVI is carried out using LS-DYNA on GFRP laminate impacted by a hemispherical striker of diameter 10 mm. Since in real-life situations structures may not be always constrained from all sides thus in this work behavior of GFRP laminate is examined when one edge (long or short) and opposite edge (long or short) of the laminate are constrained which resemble that of the cantilever and fixed type of beams, respectively. From results, it is observed that GFRP laminate under one short edge-constrained boundary condition showed 31.8% added deflection than one long edge-constrained boundary condition but one long edge boundary condition absorbed more energy than one short edge-constrained boundary condition. In case of two edge-constrained boundary conditions, two short edge-constrained boundary conditions showed partial deflection and partial penetration of the impactor while complete penetration of the striker is observed for two long edge-constrained boundary conditions without any deflection and it absorbed more energy by undergoing damage than two short edge-constrained boundary conditions.

Keywords FRP · LVI · Edge-constrained boundary condition

10.1 Introduction

During the World War II FRP composites found their application in aircraft structures. Then onwards its applicability expanded to various industries such as aerospace, automobile, marine, etc., due to its high strength to weight ratio which ultimately provides the structure less weight without any compromise in strength. These materials are

Mahesh · K. K. Singh (✉)
Indian Institute of Technology (ISM), Dhanbad Jharkhand 826004, India
e-mail: itsshindhe@gmail.com

I. Singh et al. (eds.), *Trends in Materials Engineering*,
Lecture Notes on Multidisciplinary Industrial Engineering,
https://doi.org/10.1007/978-981-13-9016-6_10

orthotropic in nature and highly vulnerable to low-velocity impact (LVI) loading, because LVI induces barely visible impact damage (BVID) into the laminate which may go unnoticed. BVID consists of subsurface micro-cracks which grow over a period of time inside the FRP structure and leads to sudden catastrophic failure of the structure. Understanding of damage mechanism associated with it becomes highly complex and difficult due to its anisotropic nature because it involves various failure mechanisms such as fiber breakage, matrix cracking, delamination, and debonding. Hence, along with experimental work, numerical analysis of FRP laminate provides a better insight in understanding the damage mechanism of FRP laminates under LVI.

Broad literature review was carried out by [1–3] covering various aspects of the impact parameters, FRP laminate parameter and numerical FEA tools. Bandaru et al. [4] conducted an experimental and numerical analysis of low-velocity impact test on Kevlar/basalt hybrid laminate with hemispherical impactor of diameter of 12.7 mm and used linear orthotropic damage model. From observation, it was found that the model used performed better in predicting the impact response for both pure and hybrid laminates. Li et al. [5] developed a numerical model for continuous fiber laminates under low-velocity impact using 9-node Lagrangian element with all edge-constrained boundary condition. Nand Kishore et al. [6] investigated the behavior of asymmetric CFRP laminate doped with MWCNTs under low-velocity impact with fully constrained circular boundary condition and observed that 2 wt% MWCNT doped CFRP laminate resulted in 13.53% more energy absorption capacity than other CFRP laminate. Giovanni Belingardin et al. [7] conducted a low-velocity impact test on GFRP laminates using drop weight impact tower with fully clamped circular boundary condition. Zuleyha et al. [8] conducted the low-velocity impact to study the interdependency between the specimen length and edge boundary conditions for GFRP cross-ply laminate. Jiang et al. [9] developed a new modeling technique for unidirectional FRP laminates under low-velocity impact loading considering the strain rate using ABAQUS in which corners of the laminate was constrained. Rawat et al. [10] examined the damage behavior CFRP laminate doped with MWCNTs under low-velocity impact with fully constrained circular boundary condition and found 0.25 wt% yielded better damage resistance properties. Balasubramani et al. [11] studied the laminate behavior under low-velocity impact using various boundary conditions, laminate thickness, striker mass, impactor velocity, and ply layup. Authors considered the following boundary conditions in their work and were clamped-clamped-clamped-clamped, clamped-simply supported-clamped-simply supported, clamped-free-clamped-free, simply supported-free-simply supported-free, clamped-free-free-free and clamped-simply supported-free-simply supported. Rawat et al. [12] used the circular boundary condition to investigate the symmetric and asymmetric laminate behavior with and without pre-crack under low-velocity impact using LS-DYNA FEA tool and found that without pre-crack symmetric laminate performed better than others while with pre-crack asymmetric laminate performed poorly when compared to other laminate. Shi et al. [13] developed a stress based damage model to study the low-velocity impact of FRP laminate and considered the fully constrained boundary condition. Volnei et al. [14] investigated the behavior of thin ply laminates under low-velocity impact by both experimental and numerical work using fully constrained circular boundary condition. Rawat et al. [15] investigated the effect of striker shape on impact damage of GFRP laminate using fully constrained boundary condition and concluded that more is the area of contact with laminate more is the damage induced.

In this numerical simulation, LS-DYNA FEA tool is used to investigate the influence of edge-constrained boundary conditions on GFRP laminate under LVI using a hemispherical impactor. Two edge-constrained boundary situations are considered in this study. Here, one short and one long edge-constrained boundary conditions are considered which fall in cantilever beam category and both short and long edge-constrained boundary conditions fall in fixed beam category. Under these boundary condition GFRP laminate damage and deflections are studied.

10.2 Materials Used

Bidirectional plain woven glass fiber is reinforced with bisphenol A epoxy resin is considered. Laminate contained eight plies with [(0, 90)/(+45, −45)/(+45, −45)/(0, 90)]S stacking sequence. Table 10.1 gives the properties of bidirectional plain woven glass fiber reinforced with bisphenol A epoxy and a striker made of mild steel of mass 20 kg is impacted onto GFRP laminate with 2 m/s velocity.

10.3 Numerical Simulation

GFRP laminate dimensions considered for numerical simulation are 150 * 100 * 4 (Length * Width * Thickness) in mm while diameter of the hemispherical striker is 10 mm with mass 20 kg impacted with 2 m/s velocity. Rigid body and solid composite failure model material cards are used to model the impactor and GFRP laminate, respectively. COMPOSITE_FAILURE_SOLID_MODEL is used to model the GFRP laminate which is based on Hashin's failure criteria, i.e., it includes five

Table 10.1 Properties of material used [15]

Laminate properties														
Young's modulus (GPa)			Poisson's ratio			Tensile strength (GPa)			Compressive strength (GPa)			Shear strength (GPa)		
E_{11}	E_{22}	E_{33}	ν_{21}	ν_{31}	ν_{32}	T_X	T_Y	T_Z	C_X	C_Y	C_Z	S_{ba}	S_{ca}	S_{cb}
26	26	8	0.1	0.25	0.25	0.85	0.85	0.12	0.72	0.72	0.5	0.105	0.065	0.065

Impactor properties	
Young's modulus (GPA)	Poisson's ration
210	0.3

failure criteria such as fiber failure in tension and compression, matrix failure in tension and compression, and shear failure of the material. Rigid_Body material card is used to model the striker. This card considers the material as a rigid body which means that the impactor undergoes the negligible amount of deformation when it interacts with the target. AUTOMATIC_SURFACR_TO_SRFACE is used to define the contact between the laminate and impactor during the course of impact and TIED _SURFACEE_TO_ SURFACE is used to define the contact in between the plies.

10.4 Results and Discussion

10.4.1 Constrained at One Edge

Figure 10.1 shows the maximum von Mises stresses for one-side constrained boundary condition. Since one edge of the GFRP laminate is constrained it can be considered as cantilever beam system. Near edge-constrained region and the impact site von Mises stresses generated are high in both the cases. Deflection of one short edge-constrained is more compared to that of one long edge-constrained GFRP composites, i.e., 20.3 and 15.4 mm, respectively (Fig. 10.2). This shows that as the distance of the free end opposite to that of the constrained edge increases, corresponding deflection also increases. But in case of short edge-constrained GFRP laminate at initial stage of impact, laminate is curved to a small extent (Fig. 10.3) instead of deflecting downward and this deflection is observed approximately till the time $t =$ 2.25 ms then exceeding this time laminate started to deflect in downward direction that is along the impact direction without taking any curved shape. The main cause for this type of behavior is the length of the unconstrained edges because in case of long edge-constrained laminate, no such behavior is observed; moreover, length of the unconstrained edge is less compared to that of the one short edge-constrained boundary condition

From Fig. 10.4, it is observed that one short edge-constrained yielded nonlinear behavior when compared to linear behavior of one long edge-constrained. For one short edge-constrained condition and one long edge constrained condition, the reduction in velocity from 2 m/s at time $t = 3.5$ ms is 1.9494 and 1.9053 m/s, respectively. This indicates that one long edge-constrained GFRP laminate absorbed more energy than one short edge-constrained boundary condition.

10.4.2 Constrained at Both Edges

Here, both long edge-constrained laminate undergoes more damage than both short edges-constrained GFRPs (Fig. 10.5). Impactor is completely penetrated inside the

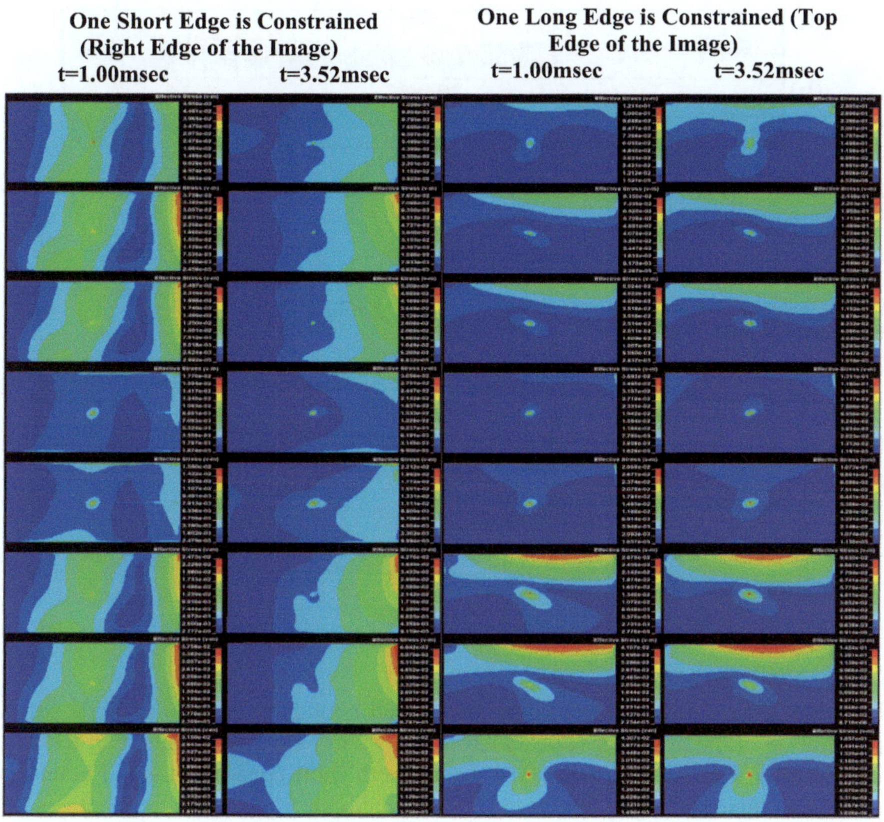

Fig. 10.1 Maximum von Mises stresses developed in one edge-constrained boundary condition

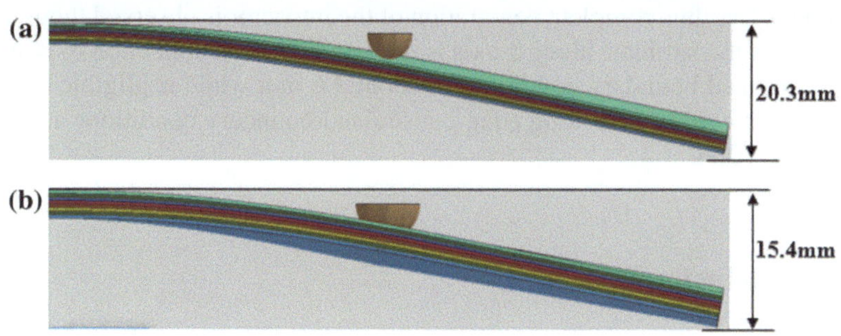

Fig. 10.2 GFRP laminate deflection at 3.52 ms, **a** one short edge-constrained, **b** one long edge-constrained

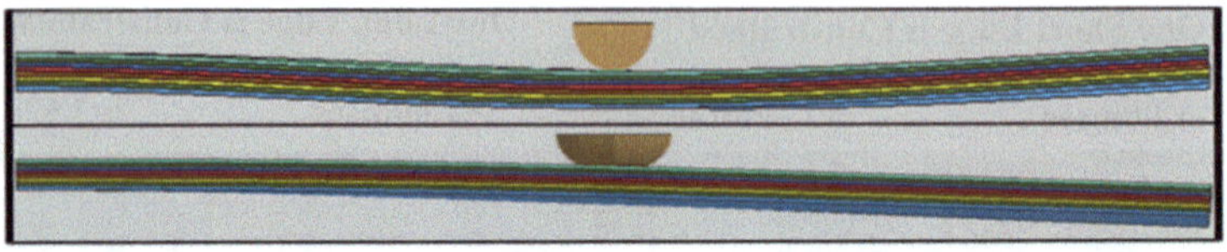

Fig. 10.3 Curved shape comparison at time $t = 1.00$ ms, **a** single short side constrained, **b** single long side constrained

Fig. 10.4 Velocity versus time of one edge-constrained boundary condition

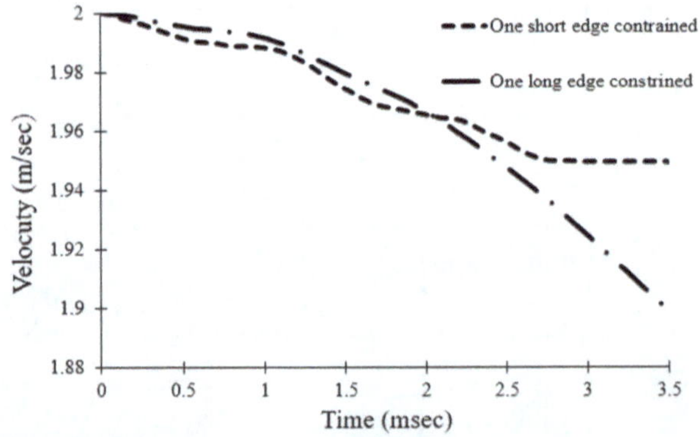

laminate in case of both long edge-constrained boundary conditions while partial penetration of the impactor is observed for both short edge-constrained GFRP laminate at time $t = 3.5$ ms (Fig. 10.6). Under both short edge-constrained boundary conditions, material region required for deformation is high; hence, laminate underwent a partial deflection and partial impactor penetration but in case of both long edge-constrained boundary conditions, material region required for deformation or deflection is less thus complete penetration of the impactor is observed though small fluctuation of the laminate along z-axis is observed. Overall deflection of both short edge-constrained boundary conditions is about 4.6 mm while negligible deflection is observed in case of both long edge-constrained boundary conditions at time $t = 3.5$ ms (Fig. 10.7).

10.5 Conclusion

LVI numerical simulation is conducted on GFRP laminate using LS-DYNA with two different kinds of boundary conditions which include one side constrained and both side constrained along short edge and long edge. These boundary conditions can be considered as cantilever beam for one edge-constrained condition and fixed type of beam for two edge-constrained conditions.

- Change in velocity for one short edge-constrained and one long edge-constrained boundary conditions is 0.065 m/s and 0.1 m/s at time $t = 3.5$ ms, respec-

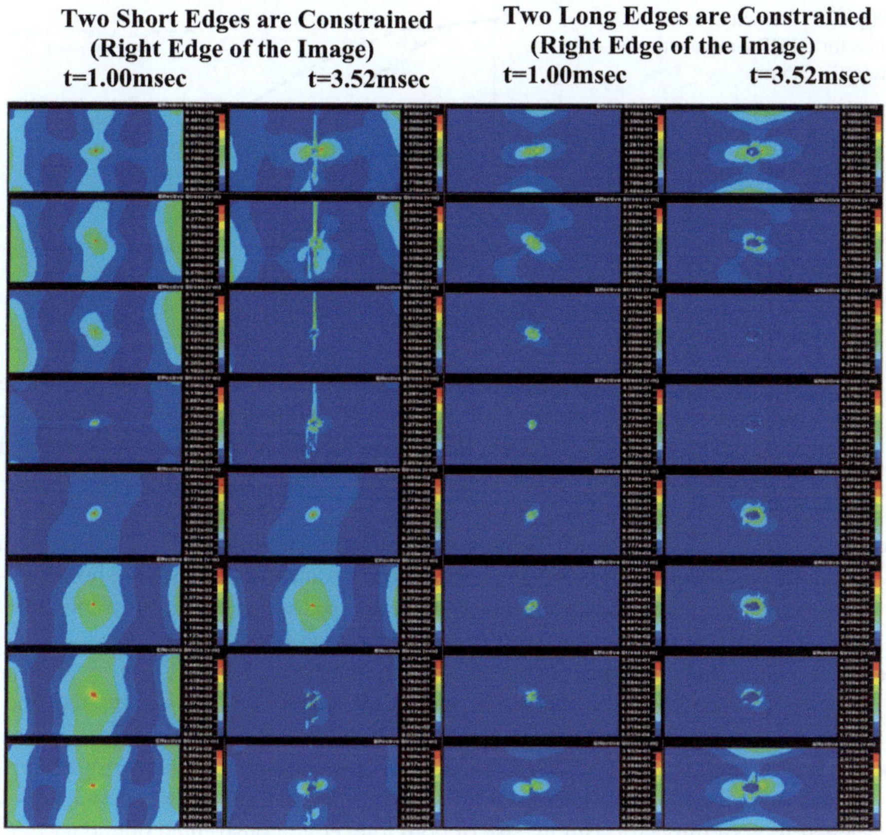

Fig. 10.5 Maximum von Mises stresses in GFRP laminate with two edge boundary conditions

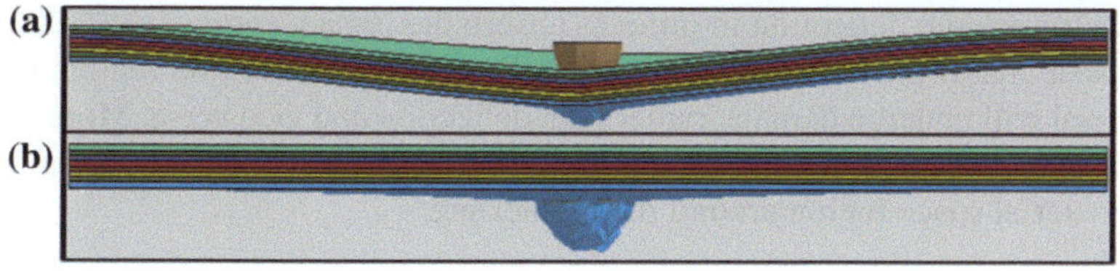

Fig. 10.6 Deformation in GFRP laminate at time $t = 3.5$ ms, **a** both short edge- constrained, **b** both long edge -constrained 1.00 ms

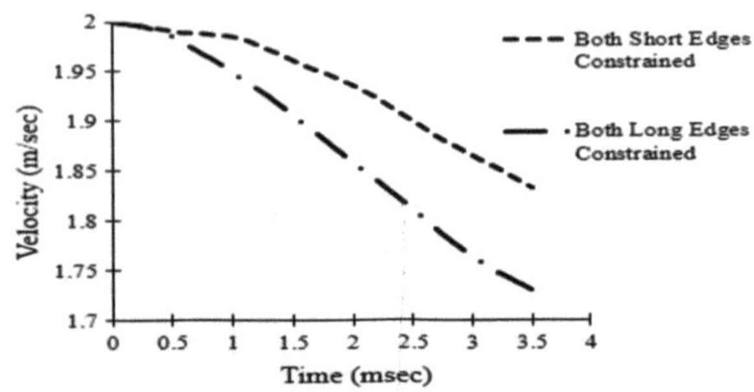

Fig. 10.7 Velocity versus time for GFRP laminate at time $t = 3.5$ ms

tively, which indicates that energy absorption is more in case of one long edge boundary-constrained condition than one short edge constrained boundary condition. Because the material space availability in case of one short edge-constrained boundary condition is more when compared to one long edge-constrained boundary condition.

- Deflection for one short edge-constrained boundary condition is 31% higher than one long edge-constrained boundary condition. Because in case of one short edge-constrained boundary condition, length of the unconstrained side is more compared that of the one long edge-constrained boundary condition.
- At the initial stage of impact in one short edge-constrained impact, energy absorption is by undergoing partial curve-shaped deflection which is a typical phenomenon observed during the numerical analysis.
- In case of both short edge-constrained boundary conditions, the energy absorption is partially due to deflection of the laminate and partially due to damage, i.e., penetration of impactor. But in case of both long edge-constrained boundary conditions, the energy absorption is completely by damage or complete penetration of the impactor is observed at the end of impact. This is due to availability of material space to undergo deflection in order to absorb the impact energy
- It is seen that if the impactor is nearer to the boundary condition region, then the material will undergo damage rather than deflection and vice versa. Moreover, the damage and deflection induced into the material are based on the availability of the material space region around the impact site.

References

1. Agrawal, S., Singh, K.K., Sarkar, P.K.: Impact damage on fiber-reinforced polymer matrix composite—a review. J. Compos. Mater. **48**(3), 317–332 (2014)
2. Singh, N.K., Singh, K.K.: Review on impact analysis of FRP composites validated by LS-DYNA. Polym. Compos. **36**, 1786–1798 (2015)
3. Rawat, P., Singh, K.K.: An impact behavior analysis of CNT-based fiber reinforced composites validated by LS-DYNA: a review. Polym. Compos. **38**, 175–184 (2017)

4. Bandaru, A.K., Patel, S., Ahmad, S., Bhatnagar, N.: An experimental and numerical investigation on the low velocity impact response of thermoplastic hybrid composites. J. Compos. Mater. **52**(7), 877–889 (2017)
5. Li, C.F., Hub, N., Yina, Y.J., Sekinec, H., Fukunaga, H.: Low-velocity impact-induced damage of continuous fiber-reinforced composite laminates. Part I: An FEM numerical model. Composites Part A **33**, 1055–1062 (2002)
6. Singh, N.K., Rawat, P., Singh, K.K.: Impact response of quasi-isotropic asymmetric carbon fabric/epoxy laminate infused with MWCNTs. Adv. Mater. Sci. Eng. (2016)
7. Belingardi, G., Vadori, R.: Low velocity impact tests of laminate glass-fiber-epoxy matrix composite material plates. Int. J. Impact Eng. **28**, 213–229 (2002)
8. Aslan, Z., Karakuzu, K., Okutan, B.: The response of laminated composite plates under low-velocity impact loading. Compos. Struct. **59**, 119–127 (2003)
9. Jiang, Z., Wen, H.M., Ren, S.L.: Modeling delamination of FRP laminates under low velocity impact. In: 3rd International Conference on Applied Materials and Manufacturing Technology, pp. 012088. IOP Science, China (2017)
10. Rawata, P., Singh K.K.: Damage tolerance of carbon fiber woven composite doped with MWCNTs under low-velocity impact. In: 11th International Symposium on Plasticity and Im-pact Mechanics, Proceedia Engineering, pp. 440–446. Elsevier, Delhi (2016)
11. Balasubramani, V., Rajendra Boopathy, S., Vasudevan, R.: Numerical analysis of low velocity impact on laminated composite plates. In: International Conference on Design and Manufacturing, Proceedia Engineering, pp. 1089–1098. Elsevier, Chennai (2013)
12. Rawat, P., Singh, K.K., Singh N.K.: Numerical investigation of low-velocity impact in symmetric and asymmetric GFRP laminate with and without pre-crack. In: Advanced Materials Proceedings, pp. 152–155. Journal of VBRI Press (2017)
13. Shi, Y., Swait, T., Soutis, C.: Modelling damage evolution in composite laminates subjected to low velocity impact. Compos. Struct. **94**, 2902–2913 (2012)
14. Tita, V., de Carvalho, Jonas, Vandepitte, D.: Failure analysis of low velocity impact on thin composite laminates: experimental and numerical approaches. Compos. Struct. **83**, 413–428 (2008)
15. Rawata, P., Singh, K.K., Singh, N.K.: Numerical investigation of damage area due to different shape of impactors at low velocity impact of GFRP laminate. In: 1st international Conference on Advancements in Aeronautical Materials for Manufacturing, Materials Today: Proceedings, pp. 8731–8738. IOP Science, Hyderabad (2017)

Chapter 11
$[ZnX4]^{2-}$ $[R]^{2+}$ Hybrid Materials: Structural and Optical Studies

Ajit Kumar, Kuldeep Kumar, Dinesh Jasrotia, Sanjay K. Verma and P. A. Alvi

Abstract The function of non-covalent interactions in structure–property relationship of $[ZnX_4]^{2-}$-based inorganic–organic hybrid materials has been studied in a series of compounds which were examined and determined with XRD data. The XRD data have been used for computer-generated structural model through computational systems to compute the non-covalent bonds in order in $ZnCl_4$, $ZnBr_4$, ZnI_4, and ZnF_4 hybrids. The molecular structures have revealed that organic materials are held inside the inorganic components through non-covalent interactions by hydrogen donor-to-acceptor, hydrogen donor to ring centroid, within the halide groups and also within the metal–metal atoms of the compound to form the different structural motifs. The selected materials were analyzed for the predominance of the hyper-Raman and Infrared spectra modes of $[ZnCl_4]^{2-}$ hybrids in comparison with the $[ZnBr_4]^{2-}$, $[ZnI_4]^{2-}[ZnF_4]^{2-}$, etc. types of hybrid derivatives are calculated through the fractional coordinates obtained from the structural data. These spectroscopic parameters indicate that such hybrid materials have optical properties.

Keywords Hybrid materials · Non-covalent interactions · Raman spectra

11.1 Introduction

The functional hybrid materials have played an important role in advanced applications of materials science in today's technocratic society. The hybrid materials are being used as energy storage devices such as in hybrid batteries and solar cells. The advanced applications of the hybrid materials in photovoltaic devices have made them very important class of materials for solar energy devices [1]. During the past

A. Kumar · K. Kumar · D. Jasrotia (✉) · S. K. Verma
Materials Research Laboratory, Department of Physics, Govt. Gandhi Memorial Science College, Cluster University of Jammu, Jammu, India
e-mail: phy.dinesh.ap@gmail.com

K. Kumar · P. A. Alvi
Department of Physics, Banasthali Vidyapith, Banasthali 304022, Rajasthan, India

I. Singh et al. (eds.), *Trends in Materials Engineering*,
Lecture Notes on Multidisciplinary Industrial Engineering,
https://doi.org/10.1007/978-981-13-9016-6_11

decade, the hybrid materials were developed on a very large scale with desired properties. The clubbing of two different components, i.e., inorganic and organic into single composite from the hybrid compound. The hybrids are classified into different types as per the mode of combining these two different moieties into single composite [2]. Class I hybrid materials are made up by the non-covalent interactions in which organic moieties are held together with the inorganic layers through the secondary interactions. The non-covalent interactions are responsible for holding the organic and inorganic moieties together into single hybrid composite. The role of these interactions is the key point to study the structural and optical behavior of hybrid materials [3]. As a part of our ongoing research on the role of non-covalent interactions in crystal physics of materials, the present study has been performed [4, 5].

11.2 Experimental Data

The hybrid materials are the class of functional materials with desired properties for the modern industrial applications, such as the energy storage devices, and photovoltaic applications which bring together the two different branches of the solid state sciences into single hybrid composite. The structural and optical studies of these materials are an important aspect of their properties and hence the applications of such materials in industry. To analyze the structural parameters of hybrids, the XRD data of 70 no. of $[ZnX_4]^{2-}$ inorganic–organic hybrid derivatives have been obtained from international data centers. The crystallographic open database has been approached to accumulate the data of $[ZnX_4]^2$ derivatives in CIF. The structural and molecular dynamics simulations to analyze the role of non-covalent interactions have been done by using the crystallographic software [6].

11.3 Discussion and Results

The analysis of hydrogen bonds calculated through XRD data by using crystallography computer simulations shows that the distance between acceptor and donor through the hydrogen atom has an array within 1.68–3.01Å and the range is 2.55–4.1Å for acceptor and donor lengths whereas the H-centered D-A angles lie in the range of 105°–175° for $[ZnCl_4]^{2-}$. Table 11.1 represents the average values of the range of hydrogen bond distances and angles. The structural analysis results that the range of H…A bond distance is 1.75–2.99(Å) for $ZnCl_4$, for $ZnBr_2$ it is 1.85–3.1(Å), and for ZnI_2. And ZnF_2 the range of bond distance is 1.95–3.0(Å) and 1.68–2.51(Å), respectively. Similarly, the range of bond angles for $ZnCl_2$, $ZnBr_2$, ZnI_2, and ZnF_2 are 105–175(°), 80–178(°), 75–165(°), and 90–175(°), respectively.

The distance versus angle graph has presented that a no. of interactions falls in the range of 1.75–3.0 Å for bond distance and 105°–175° for the bond angles through the

Table 11.1 X-H…A bonding geometry of $[ZnX_4]^{2-}$ hybrid materials

$[ZnX_4]$ hybrids	H…A(Å) range	X…A(Å) range	X-H…A(°) range
$[ZnCl_4]^{2-}$	1.75–2.99	2.55–4.1	105–175
$[ZnBr_4]^{2-}$	1.85–3.1	2.65–3.9	80–178
$[ZnI_4]^{2-}$	1.95–3.0	2.7–4.01	75–165
$[ZnF_4]^{2-}$	1.68–2.51	2.74–3.81	90–175

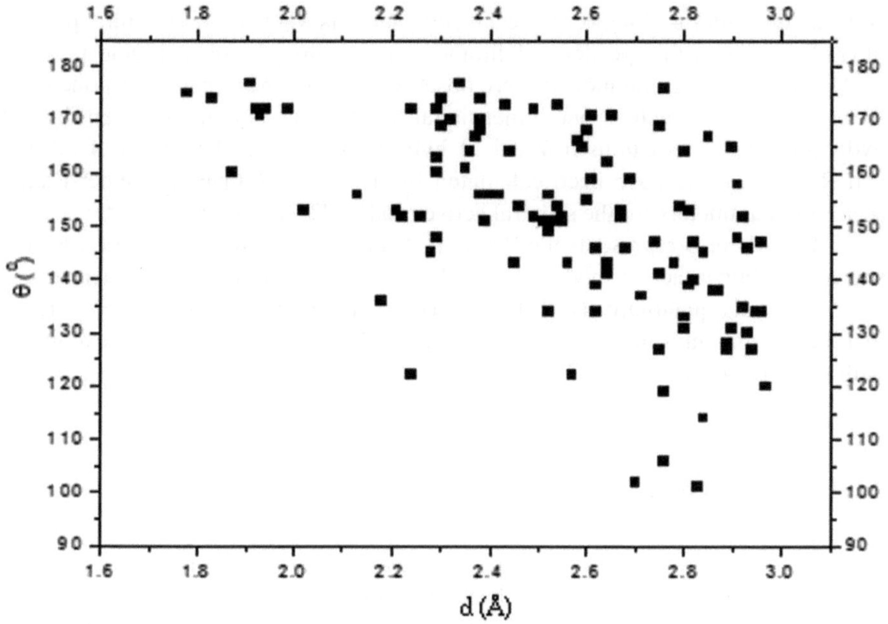

Fig. 11.1 The distance versus and angle graph for $[ZnX_4]^{2-}$ $[R]^{2+}$ hybrid materials

hydrogen bonds. In this class of hybrid materials which indicate these interactions are weak non-covalent interactions as shown in Fig. 11.1.

The ring centroid-interactions have been calculated by selecting an atom at the center of the ring and by measuring the distance of bonds and torsion angles from this point to the nearest possible combinations. The centroid-to-centroid distance for the ring conformations in $[ZnCl_4]^{2-}$ hybrids has been calculated as 3.25(1)Å in case of material no. 21 of the selected series at the symmetry code of $0.5 + x$, $0.5 + y$, z with the twisting angle of 65.45(1)°, the maximum value of centroid(π)-to-centroid(π) distance has been obtained as 3.73(1)Å in material no. 18, and the minimum value of 3.12(1)Å has been determined for the compound no. 27 of the selected series of the hybrids which results that within the inorganic patterns which are the binding forces for different structural patterns of these compounds in unit cell packing diagrams. Most of the structure of $[ZnCl_4]^{2-}$ form the 1D and 2D layered unit

Table 11.2 The centeroid–centeroid $\pi \ldots \pi$ interactions in $[ZnX_4]^{2-}$ hybrids

$[ZnX_4]^{2-}$	Center to center of ring distance in Å	Angle of twist in degrees
$[ZnCl_4]^{2-}$	3.61(7)–3.77(1)	C-π…π-C = 65.5(1)
$[ZnBr_4]^{2-}$	3.52(7)–3.78(2)	C-π…π-C = 178(1)
$[ZnI_4]^{2-}$	3.65(2)–3.79(1)	C-π…π-C = 119.1(1)

cell packing patterns along ac-plane, i.e., down *b*-axis whereas the structural patterns of $[ZnBr_4]^{2-}$ have the speeder and dimmer patterns along ac-plane (Table 11.2).

The diverse structural models were observed through the organic components by the π-interactions such as one-dimensional layers of carbon and halogen through hydrogen bond interactions in ZnCl-11 hybrid compound. The values of infrared and Raman tensors have been calculated by using the Raman and hyper-Raman scattering parameters for the spectral active modes. The equation t acoustic $= B1 + B2 + B3$, where B represents the Wyckoff's positions for IR active modes, Raman and hyper-Raman active modes [6, 7]. When we compare the crystal structure of these materials quantitatively with the theoretical structural models, it was found that these compounds have the encouraging results for Raman spectra active modes with the average values of 5.7 which depict their optical properties.

11.4 Conclusion

The analysis of non-covalent interactions such as carbon and halogen through hydrogen bond interactions in $[ZnX_4]^{2+}$-based hybrid materials shows that due to these non-covalent interactions the inorganic and organic compounds can be combined in single hybrid material. The inorganic and organic components form the different structural motifs through these non-covalent interactions. The organic moieties have the optical applications whereas the inorganic components produce the mechanical strength to these materials and carbon and halogen through hydrogen bond interactions are responsible for combining the optical as well as mechanical properties of these materials. The optical studies through these spectroscopic parameters indicate that such hybrid materials have optical properties.

References

1. Sharp, K.G.: Adv. Mater. **10**, 1243 (1998)
2. Hedrick, J.L., Magbitang, T., Connor, E.F., Glauser, T., et al.: Chem. Europ. J. **8**, 3308 (2002)
3. Kickelbick, G.: Hybrid Materials: Synthesis, Characterization and Applications. Wiley-VCH, Weinheim, Germany (2005)
4. Dinesh, J., Sanjay, K., Sridhar, B., Alvi, P.A., Ajit, K.: Mat. Chem. Phys. **207**, 98 (2018)
5. Bikram, S., Atul, T., Mukesh, K., Sanjay, K., Dinesh, J.: Mat. Chem. Phys. **206**, 34 (2017)

6. Brandenburg, K., Putz, H.: DIAMOND. Crystal Impact GbR, Bonn, Germany (1999)
7. Desiraju, G.R., Steiner, T.: The Weak Hydrogen Bond In Structural Chemistry and Biology. IUCr, UK (1999)

Chapter 12
Effect of Filler Parameters on the Mechanical Properties of Wheat Husk Filled Polystyrene Composite

Mohd. Farhan Zafar and M. Arif Siddiqui

Abstract The application of synthetic fibres in polymer composites is deteriorating because they are costly and detrimental to climate. The abundant availability of natural fibres and manufacturing ease have prompted researchers to try locally accessible natural fibres and to study their viability of reinforcement purposes and to examine their suitability as a reinforcement alternative for polymer composite. The economics and the high specific mechanical properties validate natural fibre to be a good renewable and biodegradable alternative to the synthetic reinforcement, i.e. glass fibre. Incorporation of natural fibres into the polymer is currently a standard innovation to enhance the mechanical properties of polymer. One of the biggest areas of recent development in natural fibre plastic composites worldwide is the automobile industry, where natural fibres are profitably utilized because of their low density and ever expanding climatic concerns. In this research work, the composites were synthesized using wheat husk particles as filler and polystyrene as the matrix material. Mechanical properties are studied for different filler sizes, i.e. 250–355, 355–500 and 500–710 μm and at different filler loadings, i.e. 5, 10 and 15%. Tensile strength, flexural strength, hardness and wear properties were investigated, and it was concluded that the best mechanical properties are obtained at 15% filler loading and smallest filler size studied, i.e. 250–355 μm except in case of wear rate of the composites.

Keywords Wheat husk · Polystyrene · Mechanical properties · In situ polymerization

Mohd. F. Zafar (✉) · M. A. Siddiqui
Department of Mechanical Engineering, Aligarh Muslim University, Aligarh, India
e-mail: farhan2642@gmail.com

I. Singh et al. (eds.), *Trends in Materials Engineering*,
Lecture Notes on Multidisciplinary Industrial Engineering,
https://doi.org/10.1007/978-981-13-9016-6_12

12.1 Introduction

The term "natural fibre" encompasses a broad range of vegetable, animal and mineral fibres. However, in case of composite industry, it usually refers to wood fibre and agro-based fibres like bast, leaf, seed and stem fibres. And when used in polymer composites, it provides significant reinforcement and contributes to the properties and biodegradability of the synthesized composites.

Plant fibres, for example, hemp, flax and wood, have extensive potential as support in structural materials because of the high aspect ratio and high specific strength and stiffness of the fibres [1–5]. Aside from great specific mechanical properties and positive environmental effects, various advantages for using natural fibres worth noting are minimal cost, ease of preparation, low tool wear, no skin irritation and great thermal and acoustic insulating properties [5].

Therefore, a variety of researches have done on the preparation and properties of polymer matrix composite (PMC) replacing the synthetic fibre with natural fibre like jute, sisal, pineapple, bamboo, kenaf and bagasse [6–10].

Despite the keen interest and environmental appeal of natural fibres, there use is limited to non-bearing applications due to their lower strength as compared to synthetic fibre reinforced polymer composite.

12.2 Methodology

12.2.1 Materials

Matrix Styrene monomer is employed for the preparation of the polymer composites.

Reagents Methanol was used to dissolve un-polymerized monomer during the polymerization process.

Acetone was used for washing of different glassware used during the experimentation.

Benzoyl peroxide (BPO) was used as an initiator for initiating the polymerization reaction.

Sodium hydroxide (NaOH) solution was used for washing of styrene.

Filler Mustard Husk (MH): Mustard husk is actually the outer covering of the mustard seeds which is obtained after the seeds are removed from it. This filler was obtained from agricultural fields and is cleaned for any foreign particles. The husk thus obtained is finely grounded in the food processor and segregated according to the size using standard sieves. Different filler sizes used in this study are 250–355, 355–500 and 500–710 μ. Different filler loadings used for the preparation of the composites are 5, 10 and 15% by weight of the matrix.

12.2.2 Composites Preparation

Preparation of polymer composites Styrene monomer is washed in order to remove the inhibitors present. The styrene monomer is poured into the separating funnel and an equivalent amount of 10% NaOH solution is added for washing the styrene. The separating funnel is shaken briskly so that proper washing of the monomer takes place and the inhibitor may get dissolved with the NaOH solution properly. The mixture of NaOH and inhibitor is tapped out and the washed styrene is collected in the beaker for preparation of the composites. The monomer thus collected is placed on the hot water bath which is kept at about 90–110 °C. For polymerization reaction to start, a calculated quantity of initiator (benzoyl peroxide) is added. As soon as the polymerization reaction is started, the pre-determined quantity of filler particles is added. For homogenous mixing of the filler particles, a mechanical stirrer is used for continually stirring the monomer so that no filler aggregates may be produced in the composite matrix and a more homogenous composite is obtained.

The polymer composites thus synthesized when thickens to form a paste and sticks to the beaker are poured from beaker to a plate that has been previously lined with some lubricant to prevent composite from sticking to it. To completely polymerize the collected composite, the polymer is placed in the oven for 3–4 h at 100–120 °C. The composite is now a solid mass when removed from the oven. The solid composite is broken into smaller pieces and then different test specimens were made using injection moulding.

12.2.3 Mechanical Testing

Tensile Test Tensile test was performed on universal tensile tester (UTM-G-410B) at a dwell speed of 2 mm/min. The size of the specimen was prepared according to ASTM D 638 "Type I". An average of two readings for specimens was taken for each composite sample.

Flexural Test ASTM D790 for flexural/bending test specimen. Three-point loading system was used to get the result.

Hardness test Digital Rockwell hardness tester was used for analysis of hardness test. M-Scale was used in Rockwell hardness testing having ¼ inch ball indenter. The load of 100 kg was used.

Wear Test Pin-on-disc test apparatus is employed to investigate the dry sliding wear characteristics of the synthesized polystyrene composite specimens. Composites were pressed against a rotating EN32 steel disc (hardness 65 HRC) of diameter 80 mm, and a load of 3 kg was applied normal to the specimen at a room temperature about 25 °C. The speed of the rotating disc, i.e. 500 rpm was kept constant for the entire investigation. The time for rotation for each sample was set for 5 min. The weight loss during the wear test was calculated by taking the difference in weight of the sample before and after each test.

12.3 Results and Discussion

12.3.1 Tensile Strength

The tensile strength for synthesized unfilled polystyrene is found to be 8.15 N/mm^2.

It can be observed from the tabulated results that the best tensile properties are obtained for the polystyrene composites having the highest filler loading and the smallest filler particle size, i.e. composites having 15% loading of the filler particle size of 250–355 μm. This enhancement in the tensile strength of the synthesized composite is due to the good wettability of the smaller size particle and better adhesion between the filler particles and the matrix.

It is evident from the tabulated experimental results that for the same size of filler particles, there is an increase in tensile strength at different filler loading. This increase in tensile strength is more prominent at smaller sizes of the filler particles.

As the size of the filler particles is increased, the percentage increase in the tensile strength is less abrupt as the loading is increased. It can be observed from the results that there is an increase in tensile strength for all filler loading as the filler size is increased but this increase is more noticeable in case of smaller size particle at highest level of loading (15%). The increase in tensile strength for smaller filler size is due to the fact that as the amount of filler is increased, the dispersion of filler particle increases. Also, the wettability of the filler particles increases for smaller particles at higher loading (Fig. 12.1).

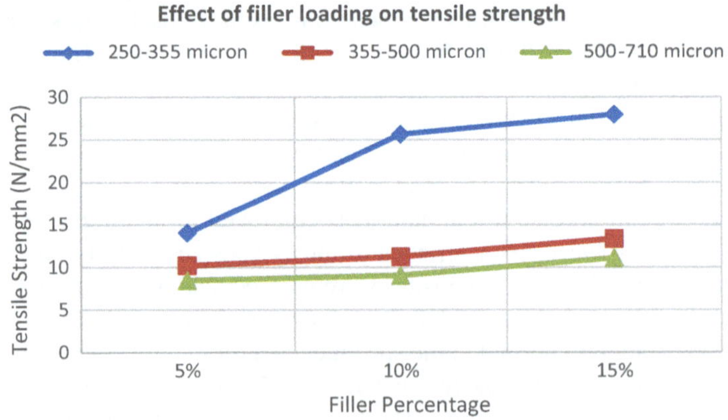

Fig. 12.1 Tensile strength at different filler compositions and sizes

12.3.2 Flexural Strength

Flexural strength for the synthesized virgin polystyrene was found to be 25.598 N/mm^2.

As discussed in case of tensile strength, the flexural strength shows the similar trend, i.e. an increase in the flexural strength is observed with an increase in fibre loading and decrease in the flexural strength as the fibre size is increased. As observed in case of tensile strength of the wheat husk filled polystyrene composites, the flexural strength tends to fall below the value of unfilled polystyrene composite at the largest filler size, i.e. 500–710 μm and filer loading of 5 and 10%. This decrease in flexural strength may be attributed to the poor interaction between the filler particles which is due to lesser wettability of larger filler particles. Also, agglomeration of the larger sized particle leads to this decrease in flexural strength.

It can be observed from the tabulated data given below that an increase in flexural strength is observed as the filler loading is increased for all particle sizes studied. Also, for any filler loading, the flexural strength tends to decrease as the filler size increases. As discussed earlier, the increase in flexural strength with increase in filler loading is due to better dispersion of filler particles at higher filler loading (Fig. 12.2).

12.3.3 Rockwell Hardness

When the experimental values of the wheat husk filled composites are observed carefully, it can be deduced that at 15% loading and the smallest filler size, i.e. 250–355 μm, the hardness value comes very close to the observed values for virgin polystyrene. The decrease in the hardness that is observed for the synthesized composites is less prominent at higher filler loading and lower filler sizes. It can be seen that at 5% filler loading, the decrease in the hardness is of the order of 32–42% for

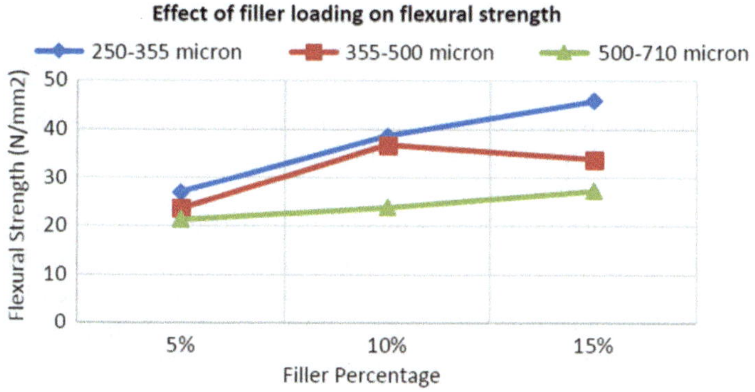

Fig. 12.2 Flexural strength at different filler compositions and sizes

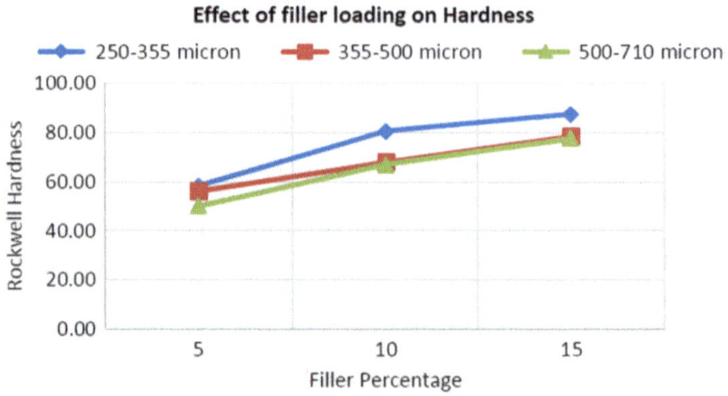

Fig. 12.3 Rockwell hardness at different filler compositions and sizes

three different filler sizes as compared to virgin polystyrene. At 10% filler loading for the smallest fibre size, i.e. 250–355 μm, the decrease in the hardness is only 6.5% while for the same filler size this decrease in hardness value converges to only 1.56% for 15% fibre loading. It can therefore be concluded that at lower sizes and higher fibre loading, the composites have shown to exhibit better hardness properties. When we examine the hardness of the composite at the highest filler loading, i.e. 15% filler loading, the decrease in the hardness values is not as prominent as observed in other filler loading, it is observed to be around 8–10% as compared to the virgin polystyrene. It can be deduced from the results that the best hardness properties were obtained for 15% filler loading and filler size of 250–355 μm (Fig. 12.3).

12.3.4 Wear Rate

From the results, it can be observed in case of the smallest fibre size of the filler particles, i.e. 250–355 μm that as we increase the loading, the wear rate tends to decrease. This decrease in wear rate is due to enhancement of the hardness of the surface of the composites. Also, the cushioning of the softer filler particles is the reason for decrease in wear rate. This same behaviour is observed for the other sizes of the wheat husk particles. In their case, the increase in the wear rate is not as high as observed in the smaller particle size.

When we observe the result for the highest filler particle sizes, i.e. 500–710 μm, it can be observed clearly that the wear rate tends to decrease with the filler loading. This decrease in the wear rate is attributed to the proper wetting of the filler particles and also the homogenous distribution of the filler particles. At higher filler loading and size, the filler particles disperse uniformly throughout the specimen thereby resulting in lesser agglomeration giving better wear properties.

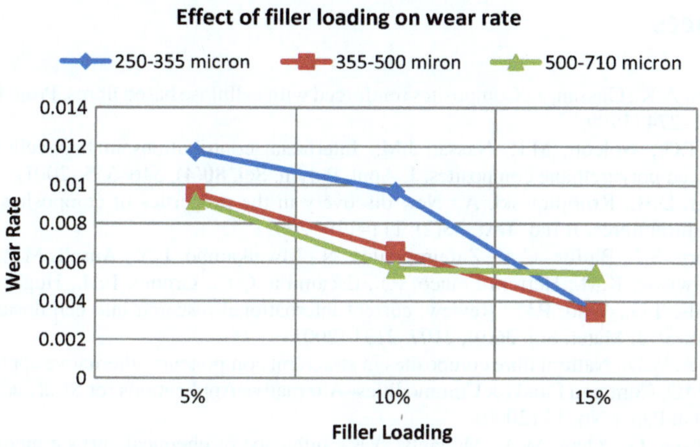

Fig. 12.4 Wear rate at different filler compositions and sizes

It can also be observed from these results that at lower filler loading, i.e. 5%, the wear rate tends to decrease with increasing filler size while at higher filler loading the wear rate tends to decrease as the filler sizes decrease. This decrease in wear rate at higher loading is due to the proper wetting of the filler particles and better filler-matrix interaction at higher loading (Fig. 12.4).

12.3.5 Conclusions

It may be concluded from the above experimental study that in situ polymerization can be successfully used for the synthesis of the composites. These composites have shown to exhibit better mechanical properties than the unfilled polymer. The tensile strength, flexural strength and hardness were found to be best in case of composites having 15% filler particles of size 250–355 μ. These mechanical properties tend to increase with an increase in filler loading and tend to decrease as the filler size increases. These findings are in conjunction with the findings from the previous researchers for the different filler-matrix combinations. [11–16]. Although in case of wear rate of the synthesized composite, it can be concluded that the smallest wear is observed in case of 15% filler loading at filler size of 500–710 μm. Also, the wheat husk filled composite can be successfully employed for various applications in the automotive industry due to its enhanced mechanical properties.

References

1. Bledzki, A.K., Gassan, J.: Composites reinforced with cellulose based fibres. Prog. Polym. Sci. **24**, 221–274 (1999)
2. Rials, T.G., Wolcott, M.P., Nassar, J.M.: Interfacial contributions in lignocellulosic fibre-reinforced polyurethane composites. J. Appl. Polym. Sci. **80**(4), 546–555 (2001)
3. Mueller, D.H., Krobjilowski, A.: New discovery in the properties of composites reinforced with natural fibres. J. Ind. Text. **33**(2), 111–129 (2003)
4. Eichhorn, S.J., Baillie, C.A., Zafeiropoulos, N., Mwaikambo, L.Y., Ansell, M.P., Dufresne, A., Entwistle, K.M., Herrera-Franco, P.J., Escamilla, G.C., Groom, L.H., Hughes, M., Hill, C., Rials, T.G., Wild, P.M.: Review: current international research into cellulosic fibres and composites. J. Mater. Sci. **36**(9), 2107–2131 (2001)
5. Brouwer, W.D.: Natural fibre composites in structural components: alternative applications for sisal, FAO, Common Fund for Commodities–Alternative Applications for Sisal and Henequen. Technical Paper No. 14 (2000)
6. Hinrichsen, G., Khan, M.A., Mohanty, A.K.: Influence of chemical surface modification on the properties of biodegradable jute fabrics—polyester amide composites. Composites: Part A **31**, 143–150 (2000)
7. Joseph, P.V., Kuruvilla, J., Thomas, S.: Effect of processing variables on the mechanical properties of sisal-fiber reinforced polypropylene composites. Compos. Sci. Technol. **59**(11), 1625–1640 (1999)
8. Mukherjee, P.S., Satyanarayana, K.G.: Structure and properties of some vegetable fibres-II. Pineapple leaf fibre. J. Mater. Sci. **21**, 51–56 (1986)
9. Hirao, K., Inagaki, H., Nakamae, K., Kotera, M., Nishino, T.K.: Kenaf reinforced biodegradable composite. Compos. Sci. Technol. **63**, 1281–1286 (2003)
10. Vazquez, A., Dominguez, V.A., Kenny, J.M.: Bagasse fibre-polypropylene based composites. J. Thermoplast. Compos. Mater. **12**(6), 477–497 (1999)
11. Atluri, R.P.V., Rao, K.M., Gupta, V.S.S.K.S.: Experimental Investigations of Mechanical properties of Golden cane fiber reinforced Polyester Composite. Int. J. Polym. Anal. Charact. **18**, 30–39 (2013)
12. Sakthivei, M., Ramesh, S.: Mechanical properties of the natural fibre (Banana, Coir, Sisal) polymer composites. Sci. Park **1**(1), 1–6 (2013)
13. Onuegbu, G.C., Igwe, I.O.: The effect of filler contents and particle size on the mechanical and end-use properties of snail shell powder filled polypropylene. Mater. Sci. Appl. **2**, 811–817 (2011)
14. Jagadeesh, D., Sudhakara, P., Lee, D.W., Kim, H.S., Kim, B.S., Song, J.I.: Mechanical properties of corn husk flour/PP bio-composites. Compos. Res. **26**(4), 213–217 (2013)
15. Hardinnawirda, K., Aisha, I.: SitiRabiatull: effect of rice husks as filler in polymer matrix composites. J. Mech. Eng. Sci. **2**, 181–186 (2012)
16. Sapuan, S.M., Bachtiar, D.: Mechanical properties of sugar palm fibre reinforced high impact polystyrene composites. Procedia Chem. **4**, 101–106 (2012)

Chapter 13
Investigation of Mechanical Properties of A6082-ZrO2-SiC Hybrid Composite by Stir and Squeeze Casting

K. Sekar

Abstract A6082-ZrO_2-SiC hybrid metal matrix composite successfully fabricated by the combined stir and squeeze casting method. The fine grain structure is observed in the optical microstructure of base alloy and hybrid composites. SEM images of A6082 with 1 wt% of SiC and 1 wt% ZrO_2 composites observed with uniform distribution of reinforcement particles. Hardness 18.6% increase in hybrid composite reinforced with 1 wt% ZrO_2 compared to the base alloy. Izod impact strength 18% increased with 1 wt% addition of ZrO_2 compared to the base alloy. Compression strength increased up to 20% in A6082 reinforced with 1 wt% ZrO_2 and SiC 1 wt%.

Keywords Stir-squeeze casting method · Hybrid composites A6082-SiC-ZrO_2 · Izod test · Compression strength · SEM images

13.1 Introduction

Aluminum-based metal matrix composite constitutes an important design class and lightweight structural materials for engineering application and mainly in aerospace and automotive applications. Fabricated Al–Si matrix reinforced with carbon fibers at varying squeeze pressures, die temperatures, and pouring temperatures. The lower die temperatures, lower pouring temperatures, and higher pressures lead to increasing the tensile strength of the composites with better bonding, decreased fiber degradation and shrinkage defects [1]. The optimum pressure squeeze casting conditions for fabricating the nanocrystalline ZrO_2/Al diameter of ZrO_2 powder 25 nm, powder preheat temperature 590–630 °C, molten A357 alloy heated to 650–750 °C, pressure 60–80 Mpa [2]. The reinforcement of ZrO_2, Al_2O_3 nanoparticles content, decreased density while yield strength, tensile strength, and compressive strength increased by 20% of the base material. The ZrO_2 reinforcement increased by 1 wt% of hardness,

K. Sekar (✉)
Department of Mechanical Engineering, National Institute of Technology Calicut, Kerala 673601, India
e-mail: sekar@nitc.ac.in

I. Singh et al. (eds.), *Trends in Materials Engineering*,
Lecture Notes on Multidisciplinary Industrial Engineering,
https://doi.org/10.1007/978-981-13-9016-6_13

but the hardness of the specimen containing 2 and 1.5 wt% ZrO_2 decreased. Due to the heterogeneous distribution of nanoparticles and high porosity content [3]. A6061-SiC composite by using stir casting the ultimate tensile strength of A6061-15 wt% SiC composite is 31.5% more than base A6061 alloy. Elongation decreased by 22% in A6061-SiC composite compared to base A6061 casting [4]. The fracture of Aluminum 6061–TiO_2 composites was ductile. With fractography, the failure of the composites consists of transgranular fracture of the TiO_2 reinforcement and ductile rupture of the Al6061 alloy matrix [5]. The pure Al with hybrid Al + Al_2O_3 + SiC produced by squeeze casting. The dispersion of Al_2O_3 and SiC particles together in an aluminum matrix improves the 67.16% impact strength and 145.51% hardness of the composite [6]. The addition of reinforcement improved the mechanical properties, when compared with base alloy [8].

13.2 Experimental Details

13.2.1 Materials

A6082 selected as matrix material, micro-silicon carbide (SiC), zirconium dioxide (ZrO_2) particles selected as reinforcement material (Table 13.1).

13.2.2 Stir Casting

A6082-ZrO_2-SiC composites were processed by combined squeeze and stir casting machine as shown in Fig. 13.2. To enhance wettability ZrO_2, SiC reinforcement is preheated at 720 °C for 1 h. The A6082 alloys are melted in furnace and maintained at 800 °C, and K_2TiF_6 is added to remove slags from molten A6082. The preheated ZrO_2, SiC reinforcement is added gradually into molten metal with the time 10 min and stirring speed of 300 rpm. After gradually mixing of reinforcement particles in molten metal, pour the molten metal into the mold. The squeeze pressure 16.89 MPa is applied to the molten metal in the die for 20s. The final specimen obtained with 46 mm diameter and 260 mm length after applying squeeze pressure as shown in Fig. 13.1 and Table 13.2.

Table 13.1 Aluminum alloy 6082 composition

Elements	Si	Mn	Fe	Cu	Mg	Cr	Ti	Zn	Al
%	1.1	0.71	0.22	0.07	0.7	0.01	0.02	0.01	Balance

Fig. 13.1 Fabricated castings

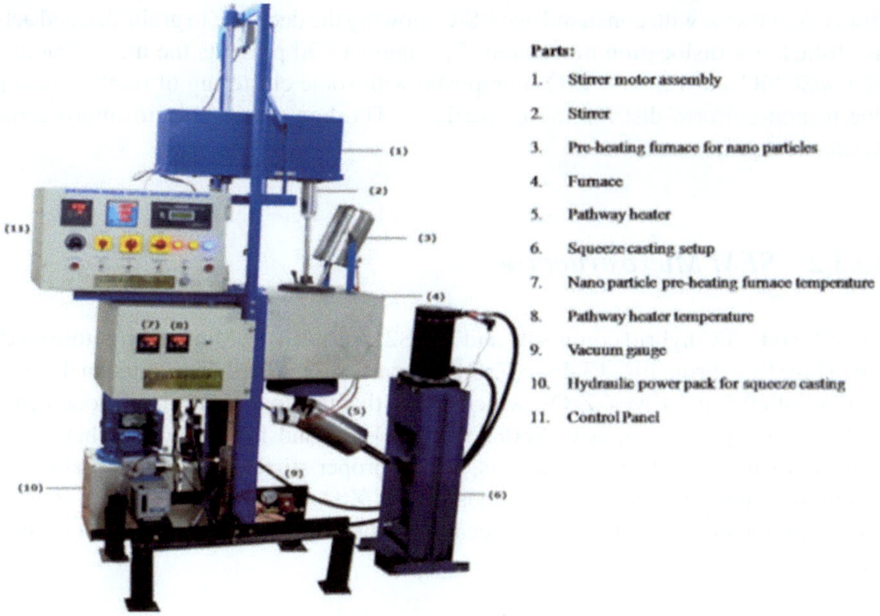

Fig. 13.2 Multiple casting machine [7]

Table 13.2 A6082 Casting alloy and composites

Cast sample No.	Castings
S1	A6082 cast alloy
S2	A6082-0.5 wt% ZrO_2-1 wt% SiC
S3	A6082-1 wt% ZrO_2-1 wt% SiC
S4	A6082-1.5 wt% ZrO_2-1 wt% SiC

13.3 Result and Discussion

13.3.1 Microstructural Studies

The microstructure of A6082 + ZrO_2 + SiC hybrid composites and A6082 alloy is characterized by using optical microscope. A6082 base material microstructure as shown in Fig. 13.3a, the grains are open and elongated due to the absence of ceramic particles. From Fig. 13.3b–c, it was observed that addition of ceramic particles ZrO_2 from 0.5 to 1 wt% with constant 1 wt% SiC showing the decrease in grain size and acts as obstacle for dislocation movement [2]. Figure 13.3d presents the microstructure of 1 wt% SiC and 1.5 wt% ZrO_2 composite with some clustering of reinforcement, due to non-uniform distribution of particles. The fine grains help to improve the mechanical properties.

13.3.2 SEM Microstructure

A6082-ZrO_2-SiC hybrid composite and A6082 base alloy SEM microstructure studied in Fig. 13.4. From Fig. 13.4b–c, ZrO_2/SiC particles are well distributed in the base matrix with 0.5 and 1 wt% ZrO_2, which shows that the composites have good dense surface. The hybrid composite A6082 with ZrO_2 0.5 and 1 wt% revealed better and uniform dispersion of reinforcements, due to proper stirring speed and preheating of reinforcement particles. The composite with ZrO_2-1.5 wt% of particles showed higher agglomeration in Fig. 13.4d because of the non-uniform distribution of reinforcement particles.

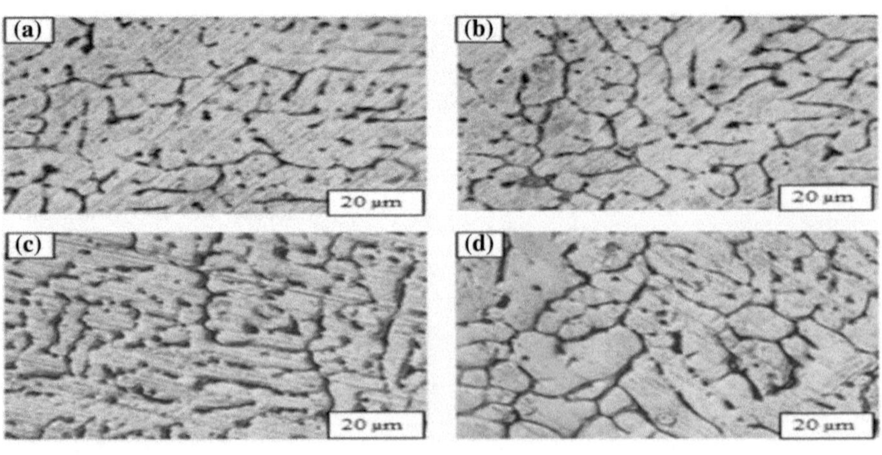

Fig. 13.3 Optical microstructure of **a** A6082 alloy, **b** A6082—0.5% ZrO_2-1% SiC, **c** A6082—1% ZrO_2—1% SiC **d** A6082—1.5% ZrO_2—1% SiC at 20 µm

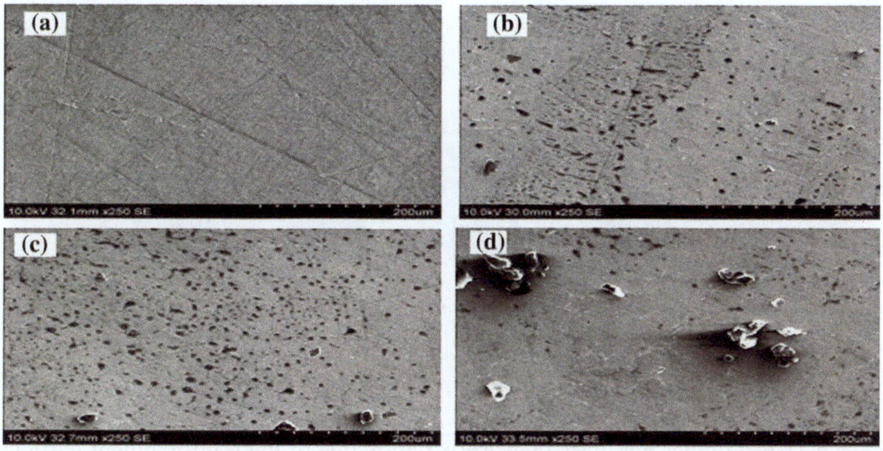

Fig. 13.4 SEM images of **a** A6082 alloy, **b** A6082—0.5% ZrO_2-1% SiC, **c** A6082—1% ZrO_2—1% SiC **d** A6082—1.5% ZrO_2—1% SiC at 200 μm

13.3.3 Hardness

According to ASTM E10-08 Vickers hardness test on A6082-ZrO_2-SiC hybrid composite and A6082 base alloy and specimens, these results are the average number of three different places of impressions. From Fig. 13.5, the hardness values measured at the various distance of 10, 20, 30, and 40 mm on the specimens.

The average hardness values of all samples compared as shown in Fig. 13.6, where all three composites showed higher hardness than the base alloy. The hardness of the hybrid composite values increased with the addition of ZrO_2 and SiC reinforcement compares to the base alloy. The hybrid composite hardness strength increased with the addition of 0.5–1 wt% ZrO_2. Further, increasing the wt% of $ZrO_{2,}$ the hardness strength is decreased because of clustering and non-uniform distribution of reinforcement particles. The composite reinforced with SiC 1 wt% and ZrO_{21} wt% has

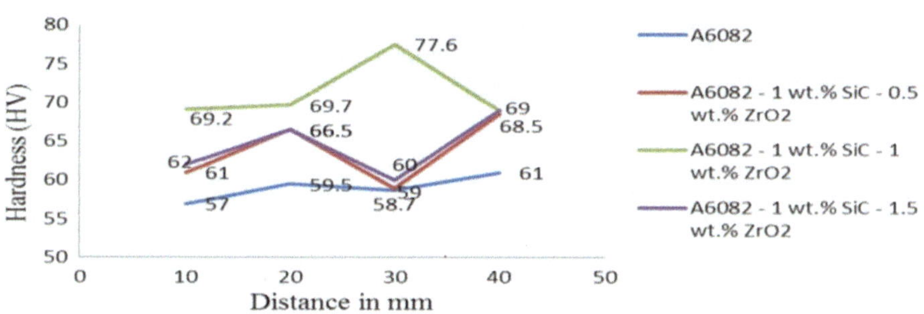

Fig. 13.5 Hardness test

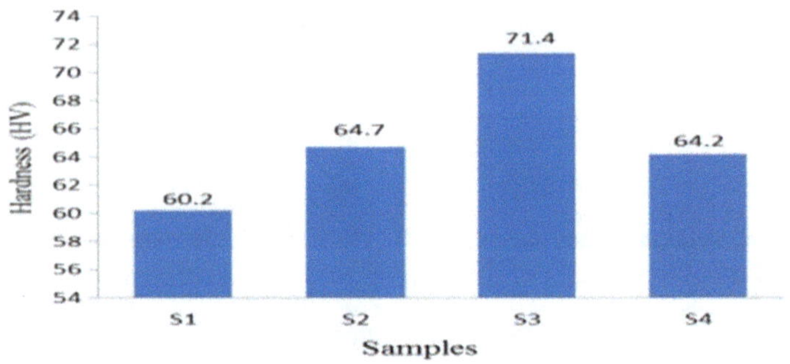

Fig. 13.6 Vickers hardness of samples

shown 18.6% higher hardness than A6082 alloy due to the reinforcement particles and their uniform distribution. Due to agglomerations and heterogeneous distribution of ceramic reinforcement particles for 1.5 wt% ZrO_2 composite, the hardness has reduced, when compared to 1 wt% ZrO_2 [4].

13.3.4 Izod Test

The Izod impact test conducted as per the ASTME23 standard (10 mm × 10 mm × 75 mm). The impact strength of A6082 base alloy and A6082-ZrO_2-SiC hybrid composite is given in Table 13.3. In Fig. 13.7, impact strength is increased with increasing SiC, ZrO_2 ceramic reinforcement particles up to 1 wt% ZrO_2 due to increased ductileness so that toughness increased [2]. The increased wt% of ZrO_2, the Izod impact strength is decreased due to the clustering effect.

13.3.5 Compression Test

The compression test performed as per ASTM E9 (20 mm × 20 mm × 20 mm) at 2 KN/Sec load rate. The stress versus strain graph and yield strength plotted based on the compression test are shown in Figs. 13.8 and 13.9.

Table 13.3 Izod impact strength

Sample	Toughness (Joules)
1	22
2	24
3	26
4	25

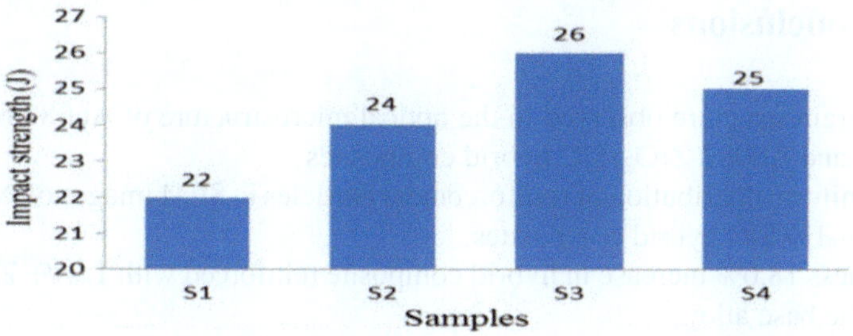

Fig. 13.7 Izod impact strength of the different sample

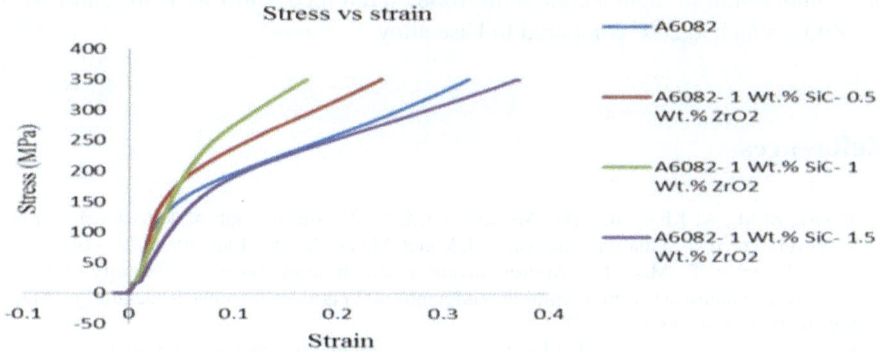

Fig. 13.8 Combined stress versus strain diagrams of various samples

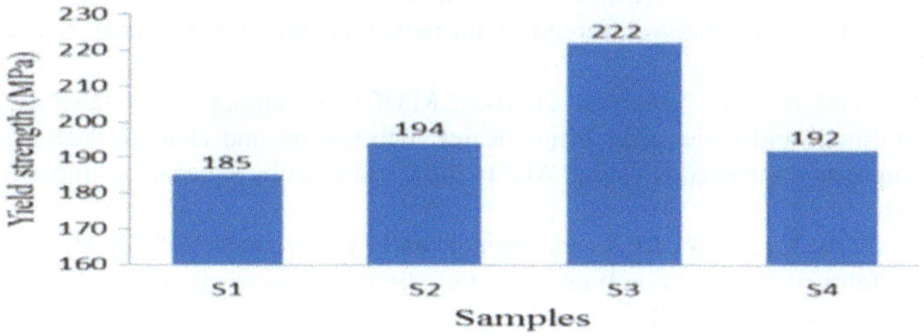

Fig. 13.9 Yield strength of various samples

The compression strength was increased up to 1 wt% of ZrO_2 that is because of the ceramic reinforcement particles uniformly distributed [1]. The compressive strength of A6082 with 1.5 wt% of ZrO_2 composite has reduced, when compared to 1 wt% of ZrO_2 composite, but higher than the base alloy. Formation of agglomeration may be the reason for this compressive strength reduction at 1.5 wt% ZrO_2 and 1 wt% SiC hybrid composite. The compressive strength of composite 1 wt% ZrO_2 and 1 wt% SiC is 20% more than the A6082 alloy.

13.4 Conclusions

1. Fine grain structure observed in the optical microstructure of A6082 base.
2. Alloy and A6082-ZrO_2-SiC hybrid composites.
3. The uniform distribution of reinforcement particles in SEM images of A6082-1% ZrO_2—1% SiC hybrid composites.
4. Hardness 18.6% increase in hybrid composite reinforced with 1 wt% ZrO_2 more than the base alloy.
5. With the addition of 1 wt% $ZrO_{2,}$ Izod impact strength of the composite increased 18% compared to the base alloy.
6. Compression strength increased in A6082 reinforced with 1 wt% SiC and 1 wt%. ZrO_2, which is 20% compared to base alloy.

References

1. Moosa, M.M., Al-Khazraji, K.K., Muhammed, O.S.: Tensile strength of squeeze cast carbon fibers reinforced al-Si matrix composites. J. Miner. Mater. Charact. Eng. **10**(2), 127–141 (2011)
2. Geng, L., Imai, T., Mao, J.F.: Microstructure and high strain rate superplasticity of in situ composite synthesised from aluminium and nanoZrO2 particles by powder metallurgy. Mater. Sci. Technol. **17** (2001)
3. Hajizamani, M., Baharvandi, H.: Fabrication and studying the mechanical properties of A356 alloy reinforced with Al2O3-10% Vol. ZrO2 nanoparticles through stir casting. Adv. Mater. Phys. Chem. **1**, 26–30 (2011)
4. Murtya, S.V.S.N., Rao, B.N., Kashyap, B.P.: On the hot working characteristics of 6061Al–SiC and 6061–Al2O3 particulate reinforced metal matrix composites. Compos. Sci. Technol. **63**, 119–135 (2003)
5. Patel, D.R., Patel, N.S.: Development of Al-Sic MMCs for making valves. IJARIIE **2**(3) (2016)
6. Gurwinder Singh and Charanjeet Singh Kalra: Fabrication and characterisation of al-based hybrid composite reinforced with Sic, Al_2O_3 and C particles by squeeze casting. Mech. Confab **3**(1) (2014)
7. Sekar, K., Allesu, K., Joseph, M.A.: Mechanical and wear properties of Al–Al_2O_3 metal matrix composites fabricated by the combined effect of stir and squeeze casting method. Trans. Indian Inst. Met. **68**(2), 115–121 (2015)
8. Sekar, K., Manohar, M., Jayakumar, K.: Mechanical and tribological properties of A356/$Al2O_3$/MoS_2 hybrid composites synthesized through combined stir and squeeze casting. Adv. Mater. Metall. Lecture Notes in Mechanical Engineering, https://doi.org/10.1007/978-981-13-1780-4_13

Chapter 14
Thermal Modeling of Beryllium Copper Alloy C-17200 for Electric Discharge Machining

Dhruv Joshi, Satyendra Singh, Chandra Prakash and Vinod Kumar

Abstract The parameters which effect die sinker electric discharge machine's material removal rate are studied over here. The material focused in this experimental study is beryllium copper alloy C-17200, and electrode/tool used for machining is made of mixture of graphite and brass on EDM model EDM4025. Material removal rate is calculated and analyzed for selected set of machining parameters. The thermal analysis plays an important role in analyzing and solving the EDM of Be–Cu in terms of heat distribution. A thermal model using simulation software ANSYS is presented in order to distinguish the influence of factors current and pulse on time in EDM process of heat circulation beside the depth and radius of machining workpiece. As beryllium copper alloy (C-17200) has many applications and because of properties that is strong oxidization resistance, high hardness and elevated wear resistance Be-Cu has extensively used in industrial application such as in dies and cutting tools, automotive and aerospace industries.

Keywords Beryllium copper alloy (C-17200) · ANSYS · Graphite electrode · Thermal modeling · Process parameters

14.1 Introduction

In an electric discharge machine, the electrical energy is utilized for generation of electric spark (generated due to elevated heat and high temperature) and material removal takes place due to thermal energy produced by this spark in between the tool and electrode. The tool is used to cut the workpiece into desired shape. Different work piece materials which have difficulties in machining due to their some properties are machined in this machining process. The shock waves produced during spark

D. Joshi (✉)
B.T.K.I.T. Dwarahat, Almora, Uttarakhand, India
e-mail: dhruv31joshi@gmail.com

S. Singh · C. Prakash · V. Kumar
Department of Mechanical Engineering, B.T.K.I.T. Dwarahat, Almora, India

I. Singh et al. (eds.), *Trends in Materials Engineering*,
Lecture Notes on Multidisciplinary Industrial Engineering,
https://doi.org/10.1007/978-981-13-9016-6_14

are used for production of potential difference [1] and thus for removal of material. The material elimination process takes place by sequence of spark for a time span in presence of dielectric fluid between tool and workpiece. A small distance (gap) between the electrodes is maintained (tool and workpiece) with no mechanical contact thus successively remove a volume of workpiece material and is carried away from gap by the dielectric flow in debris form.

14.2 Literature Survey

Thermal model for performance parameter's surface roughness and material removal has been developed [2]. These factors' experimental values are determined, considering discrete set of input parameters. The effect of process parameter on MRR of beryllium copper by the use of full factorial technique in sinking EDM was investigated [3] and resulted that current was the most considerable process parameter affecting MRR and voltage shows least influence for removing material while machining. The temperature distribution was experimentally determined using hyperbolic heat conduction model in EDM [4]. Thermal electrical model of EDM was investigated using joule heating factor, and maximum temperature gained during machining was found. The parameter's material removal rate and surface roughness were estimated for AlSi, obtaining maximum temperature in the discharge feed [5]. The parameter current is the main factor for the improvement of MRR as well as the roughness.

14.3 Process Modeling and Theoretical Analysis

The EDM machining is a complex process which involves mainly modes of heat transfer. The major mode throughout the workpiece is heat conduction, and the other mode occurred between workpiece and dielectric fluid is convection. The systematic diagram of sinker EDM is shown in Fig. 14.1 [6].

EDM is an approximate process because of the ambiguity arising due to factors such as shape of plasma channel, radius of plasma, source of heat, circulation of energy during machining and physical and chemical properties of the material. These factors contribute to the precise forecast of EDM models [4]. The following assumptions are considered in the present analysis for defining problem mathematically feasible.

14.3.1 Assumptions

- The modeling and analysis represent result for one single spark.
- Thermal properties of this workpiece are dependent on temperature.

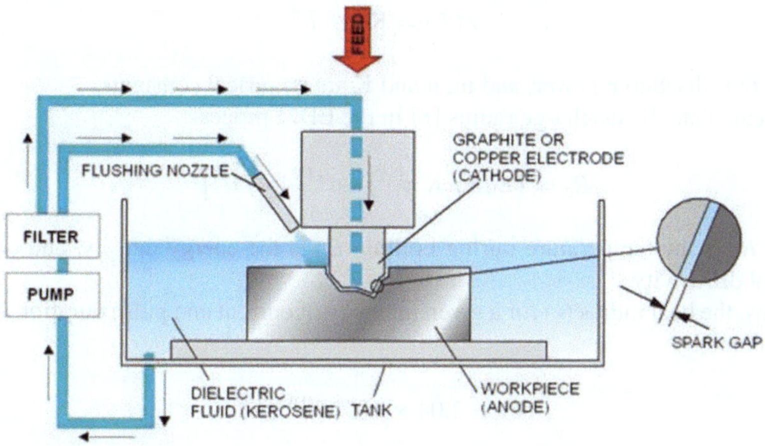

Fig. 14.1 Schematic diagram of EDM [6]

- The workpiece shape is not affected by the heat generation during machining because during thermal heating the expansion of the body is negligible.
- The effect of latent heat of fusion and vaporization is considered negligible.
- Temperature independent properties are density and specific heat.

14.3.2 Governing Equation

The governing equation of heat conduction in axi-symmetric [2], nonlinear and transient model can be attained by the Fourier equation of heat conduction and is as follows:

$$\frac{1}{r}\frac{\partial}{\partial r}\left(kr\frac{\partial T}{\partial r}\right)+\frac{\partial}{\partial z}\left(k\frac{\partial t}{\partial z}\right)=\rho c_{\mathrm{p}}\frac{\partial T}{\partial t} \quad (1)$$

where ρ denotes density $\frac{\mathrm{kg}}{\mathrm{m}^3}$, c_{p} denotes specific heat $\frac{\mathrm{J}}{\mathrm{kg\cdot K}}$, k denotes thermal conductivity of the workpiece $\frac{\mathrm{W}}{\mathrm{m\cdot K}}$, T denotes temperature (K), t denotes the time, and r and z denotes coordinates of the workpiece.

14.3.3 Heat Flux and Spark Radius

The below Equation [7] is discharge power and time dependent and has been used to calculate the radius in EDM process,

$$r(T) = \mathrm{K}\, Q^{\mathrm{m}}\, T^{\mathrm{n}} \tag{2}$$

where Q is discharge power, and m, n and K are empirical constants.

To calculate the discharge radius [8] in the EDM process:

$$R_{\mathrm{b}} = \left(E_0 R / K \pi^{0.5}\right) \tan^{-1}\left[4\alpha t / R^2\right]^{0.5} \tag{3}$$

where R_{b} is the temperature during boiling, E_0 is the energy density, and α is the thermal diffusivity.

Thus, the heat radius (r) for a given intensity of current and pulse duration [7] can be obtained by:

$$(r) = 2.04 \times 10^{-3}\, I^{0.43} t^{0.44} \tag{4}$$

where (I) is intensity of current and (t) duration of pulse.

14.4 Experimental Methodology

An EDM machine has been used for performing the experiments for the sake of making holes on workpiece material of beryllium copper alloy (C-17200). The machine consists of dielectric fluid tank, electrode guide, servo motors, work table and electrode holder. Figure 14.2 shows the pictorial view of EDM 4025.

During machining, the workpiece material was fixed on fixture. Figure 14.3 shows the workpiece and electrode arrangement of EDM of Be–Cu C-17200.

14.4.1 Electrode

Electrode with a mixture of graphite and brass is used. The diameter of solid electrode is 20 mm. and length is 800 mm. Figure 14.4 shows the electrode and workpiece material. **Brass** is a copper and zinc alloy and provides better MRR while **graphite** provides a cleaning action at low speeds.

14.5 Result and Discussion

The parameters considered for the experiment are current, voltage and pulse on time which are analyzed using ANSYS Mechanical APDL 16.0 tool for obtaining the distribution of temperature beside the depth and radius of workpiece. Here, APDL symbolizes ANSYS Design Parametric Language. The results of thermal analysis

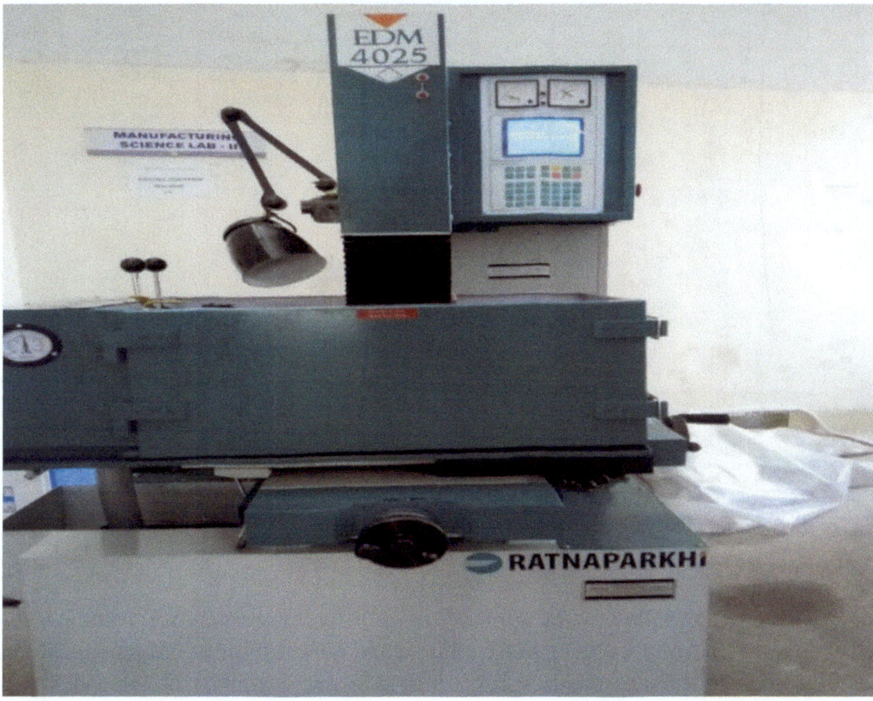

Fig. 14.2 Pictorial view of EDM-4025

Fig. 14.3 EDM of Be–Cu C-17200

Fig. 14.4 Pictorial view of electrode and workpiece material

for Be–Cu C-17200 alloy by the application of different intensity for a single spark for parameter's current, voltage, pulse on time and pulse off time within non-moving electrode, i.e., stationary position for a single spark, are given in Fig. 14.5. The modeling of the present work is done considering the base paper [9].

The MRR value was determined [7] using the equation.

$$\mathrm{MRR} = \frac{w_\mathrm{b} - w_\mathrm{a}}{t_\mathrm{m}}\ (\mathrm{g/min}) \tag{5}$$

where

w_a = Material weight before machining
w_b= Material weight after machining
t_m = Machining time

Figure 14.5a validates the contour for temperature at the pulse on time's completion, and Fig. 14.5b shows the contour of temperature while the pulse off time finish. It is casually obvious from Fig. 14.5 that at the surface, top face occurs a maximum temperature, where the intensity of heat flux is maximum as per Gaussian heat flux distribution. Also, Fig. 14.5a points toward the rise of temperature for the spark on time for melting of material, as the temperature, which is maximum, is higher than the melting point of Be–Cu C-17200.

As no plasma channel occurs among the electrode and workpiece during the off time of pulse, hence no heat flux is experienced. Convection phenomenon arises in between workpiece and dielectric fluid. It can be observed from Fig. 14.5b that by the dielectric fluid through convection, most of the heat is taken away and the reason behind this is sharp rise in temperature in on-time step.

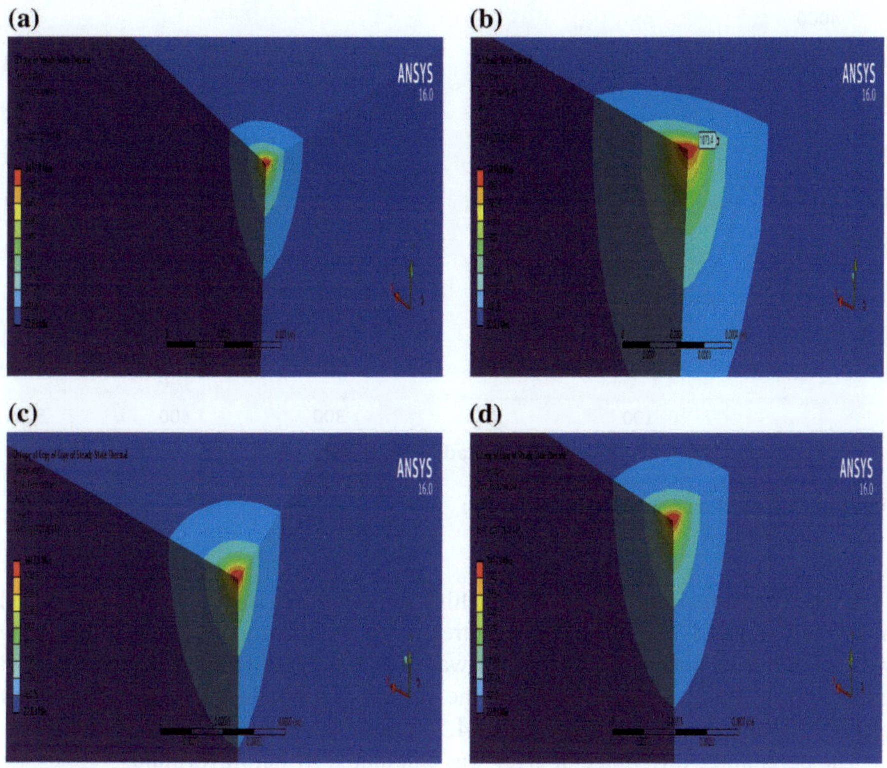

Fig. 14.5 Numerical results for Be–Cu with $I = 8$ A, $V = 40$ V, **a** Temperature contour for a single spark just at the end of pulse on time (20 μs), **b** Temperature contour for a single spark just at the end of pulse off time (30 μs), **c** Temperature distribution along workpiece radius, **d** Temperature distribution along workpiece depth

Figure 14.5c, d is indicative of the distribution of temperature in radius and depth of the workpiece, respectively. By comparing, Fig. 14.5a–d exposes that the maximum temperature during pulse off time would obviously decrease.

Also, there is a slighter decrease in temperature at the surrounding area of the workpiece center in the radius to that of depth direction. Therefore, alongside the radial direction the material removal rate is more than that of depth.

The effect of current on distribution of heat is investigated and shown in Fig. 14.6 and detected that with rapidly increasing the pulse current, the rise in distributed temperature will definitely take place with energy densities of high levels.

It is perceived that the temperature of top surface increases with current rise. This is because the current is a factor behind transfer of heat to the workpiece. Thus, larger the current, the more heat energy will be generated.

The EDM experiments were performed with the intention of studying the process parameter's effect on the output response individualities.

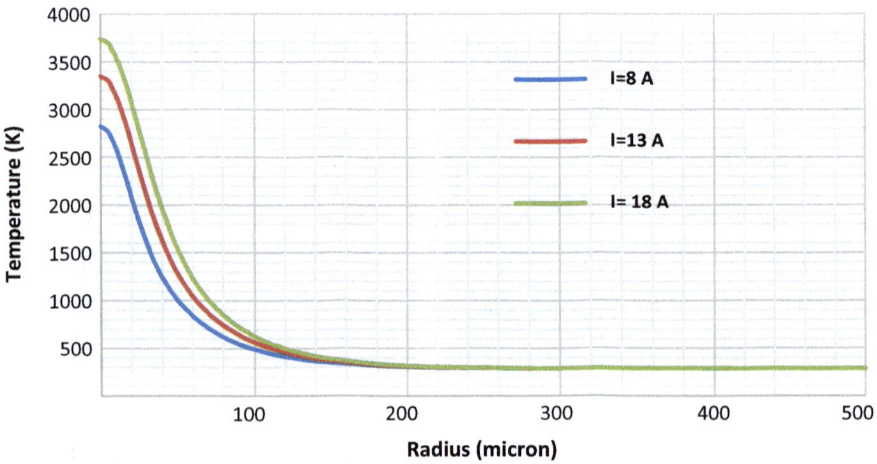

Fig. 14.6 Effect of current on heat distribution

In each experiment, the value of machining parameters was set in machine based as shown in Table 14.1. In order to measure the MRR experimentally, machining time was set and thus observed in a careful way for better results. Mathematical model generated reliable and acceptable and the variation of obtained experimental result of MRR with current is shown in Fig. 14.7.

The percentage difference or error was calculated by below formula:

$$\text{Percentage difference} = \frac{\text{Experimental}_{\text{MRR}} - \text{Analyzed}_{\text{MRR}}}{\text{Experimental}_{\text{MRR}}} \times 100$$

In each experiment, the value of machining parameters was set in machine based as shown in Table 14.1. In order to measure the MRR experimentally, machining time from starting of discharge to the end of machining is observed.

Table 14.1 Experimental data collection of MRR

S. No.	Peak current (A)	Voltage (V)	Pulse on time (μs)	Pulse off time (μs)	Machining time (min)	MRR (g/min)
1	5	35	50	100	70	0.0331
2	8	45	100	100	55	0.0359
3	10	40	102.5	145	45	0.0387
4	13	45	165	155	35	0.1159
5	18	40	180	300	55	0.1493
6	20	50	200	200	50	0.1549
7	24	55	200	100	30	0.1623

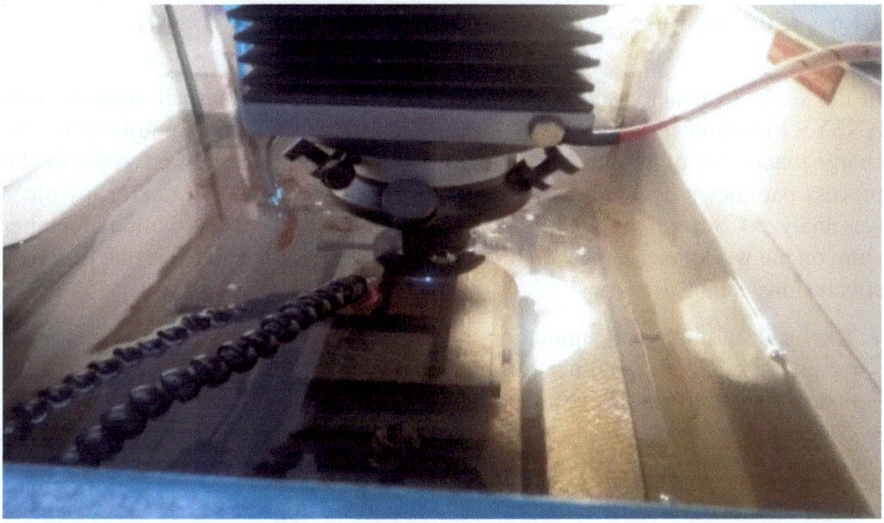

Fig. 14.7 Material cutting during machining process

Table 14.2 Comparison of experimental and analytical data collection of MRR

S.No.	Current (A)	Experimental MRR	Analyzed MRR	% Difference
1	5	0.0331	0.0245	2.5
2	8	0.0359	0.0286	2.8
3	10	0.0387	0.0379	3.7
4	13	0.1159	0.1102	4.9
5	18	0.1493	0.1413	5.3
6	20	0.1549	0.1501	3.1
7	24	0.1623	0.1596	2.9

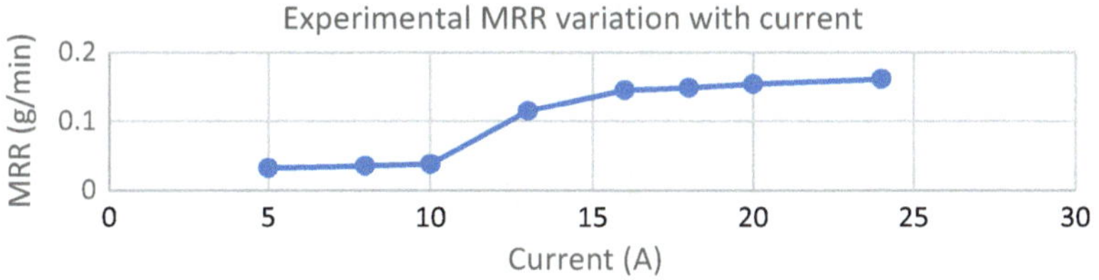

Fig. 14.8 Experimental MRR variation with current

Table 14.2 shows the effect of machining parameters on MRR for experimental and analyzed result (using ANSYS 16.0) of MRR. Also, percentage of difference among analyzed and experimental value of MRR is calculated, and thus, it is definite that mathematical model generated reliable and acceptable and the variation of obtained experimental result of MRR with current is shown in Fig. 14.8.

14.6 Conclusion

In the study over here, an experimental investigation and thermal analysis using ANSYS Mechanical APDL (ANSYS Design Parametric Language) of electro discharge machining of beryllium copper alloy (C-17200) are performed. And effect of process parameters including voltage, pulse on time, pulse current and gap voltage was analyzed. Several set of experiment have been done, and distribution of temperature on the workpiece is measured. Based on the results and analysis, the following conclusions can be drawn:

- The result of numerical simulation indicates that current is the major influencing parameter compared to other considered parameters in thermal analysis which is in an agreement with experimental result.
- There occurs a noteworthy rise in temperature of workpiece by increasing the current. This results in high level of energy density during the machining operation.

References

1. Joshi, S.N., Pande, S.S.: Development of an intelligent process model for EDM. Int. J. Adv. Manuf. Technol. **45**, 300–317 (2009)
2. Yadav, V., Jain, V.K., Dixit, P.M.: Thermal stresses due to electrical discharge machining. Int. J. Mach. Tools Manuf. **42**, 877–888 (2002)
3. Joshi, S.N., Pande, S.S.: Thermo-physical modelling of die-sinking EDM process, J. Manuf. Process. **12**(1), 45–56 (2010)
4. Singh, A., Ghosh, A.: A thermo-electric model of material removal during electric discharge machining. Int. J. Mach. Tools Manuf. **39**, 669–682 (1999)
5. Das, S., Klotz, M., Klocke, F.: EDM simulation: finite element-based calculation of deformation, microstructure and residual stresses. J. Mater. Process. Technol. **142**, 434–451 (2003)
6. Sommer, C., Sommer, S.: Complete EDM handbook: wire EDM, RAM EDM, small hole EDM: practical information for designer, engineers, machinists, tool and die makers, mold makers and others in the metal machining fields. Advance Publications (2005)
7. Pandey, P.C., Jilani, S.T.: Plasma channel growth and the resolidified layer in EDM. Precis. Eng. **8**(2), 104–110 (1986)
8. Pant, M., Radha, T., Singh, V.P.: A simple diversity guided particle swarm optimization. In: *IEE Congress on Evolutionary Computation—2007*, IEEE Explore, pp. 3294–3299 (2007)
9. Mohanty, C.P., Sahu, J., Mahapatra, S.S.: Thermal-structural analysis of electrical discharge machining process. Proc. Eng. **51**, 508–513 (2013)

MIX
Papier aus verantwortungsvollen Quellen
Paper from responsible sources
FSC® C105338

If you have any concerns about our products,
you can contact us on
ProductSafety@springernature.com

In case Publisher is established outside the EU,
the EU authorized representative is:
Springer Nature Customer Service Center GmbH
Europaplatz 3, 69115 Heidelberg, Germany

Printed by Libri Plureos GmbH
in Hamburg, Germany